DONORA DEATH FOG

Frontispiece. Aerial view of Donora on October 31, 1948. Note the smoke trailing from the Zinc Works (*lower right*). Courtesy of Donora Historical Society.

DONORA DEATH FOG

CLEAN AIR AND THE TRAGEDY OF A PENNSYLVANIA MILL TOWN

ANDY McPHEE

UNIVERSITY OF PITTSBURGH PRESS

Published by the University of Pittsburgh Press, Pittsburgh, Pa., 15260
Copyright © 2023, Andy McPhee
Manufactured in the United States of America
Printed on acid-free paper
10 9 8 7 6 5 4 3 2 1

Cataloging-in-Publication data is available from the Library of Congress

ISBN 13: 978-0-8229-6671-5
ISBN 10: 0-8229-6671-9

Cover photo: The wire mill, Donora, PA, circa 1910. Photo by Bruce Dresbach. Library of Congress Prints and Photographs Division, Washington, DC, 2002713075.

Cover design: Joel W. Coggins

TO THE TWENTY-ONE PEOPLE WHO DIED IN THE SMOG, AS WELL AS to the untold number of victims who perished in the months and years that followed. Their unwitting sacrifice has given the nation increasingly cleaner and more healthful air.

IVAN CEH

BARBARA CHINCHAR

TAYLOR CIRCLE

JOHN CUNNINGHAM

BERNARDO DI SANZA

MICHAEL DORINCZ

WILLIAM GARDINER

SUSAN GNORA

MILTON HALL

EMMA HOBBS

IGNACE HOLLOWITI

GEORGE HVIZDAK

JANE KIRKWOOD

MARCEL KRASKA

ANDREW ODELGA

IDA ORR

THOMAS SHORT

PETER STARCOVICH

PERRY STEVENS

SAWKA TRUBALIS

JOHN WEST

———————

Most especially to my glorious wife, Gay: my partner, my best friend, and the forever love of my life.

Freshly charged, the zinc smelting furnaces, crawling with small flames, yellow, blue, green, filled the valley with smoke. Acrid and poisonous, worse than anything a steel mill belched forth, it penetrated everywhere . . . setting the river-boat pilots to cursing God, and destroying every living thing on the hills.

Thomas Bell, *Out of This Furnace: A Novel of Immigrant Labor in America*

CONTENTS

FOREWORD

I RELY ON THE WELL-KNOWN EXAMPLE OF DONORA, PENNSYLVANIA, IN the classes I teach on air pollution and environmental regulation. From October 26 to 31, 1948, a cloud of smoke hung over the Monongahela Valley town, trapped by an inversion and the mountainside. Although the emissions of sulfur dioxide, fluorine, and carbon monoxide gases and particulate lead and zinc were accumulating in the stagnant air mass, operations at the Donora Zinc Works and the American Steel & Wire Company continued without interruption during the six-day period. Dense smoke darkened the sky. Nearly half the town of fourteen thousand became sickened in the ensuing days, and twenty-one people died. Many current students seeing pictures of these scenes for the first time are shocked that this occurred in the United States, because they have never known a time when industrial air pollution was so substantial.

Donora, along with the London Smog of 1952, numerous fires on the Cuyahoga River in Cleveland, increasing photochemical smog in Los Angeles, and growing recognition of the severe impact of lead on children's brain development, brought the hazards of air pollution exposure into the public consciousness. In the years that followed, several pieces of air quality legislation were passed in Congress, but progress was slow due to the common sentiment that implementation of such measures should be determined by individual states rather than at the federal level. Several legislative steps offered progress, but it was not until the Clean Air Act Amendments of 1970 that a federal program for the prevention and control of air pollution was passed. Strengthened by additional amendments in 1977 and 1990, the Clean Air Act has lowered air pollutant concentrations to a fraction of 1970s levels while saving the American public thirty times the cost of the regulation. These measures have resulted in improved quality of life and reduced premature mortality rates throughout the United States.

When we think about the events of Donora and its aftermath, it is easy to focus on metrics: morbidity, mortality, concentrations, laws, regulations, and prevented future morbidity and mortality. But the disaster was also a personal tragedy for Donora residents, many of whom struggled with dual loyalty to the companies that put food on their tables and to their family members who became ill from the chemical exposures.

To this day, residents cope with Donora's tainted legacy, as grieving families received trivial compensation for their losses and the town bore the stigma of what happened. Many could not leave Donora if they wanted to, and many knew no other alternative. *Donora Death Fog* provides a comprehensive history of the events that occurred during the days of the smog along with historical context of the circumstances contributing to the tragedy. For this book, Andy McPhee painstakingly researched the personal story of each individual who died from the toxic air emissions. To date, many of those individual stories, including the emotional and financial dependence experienced by the citizenry on the companies that poisoned them, cannot be found elsewhere. In exploring the evolution of dual loyalty developed by community members during the half century prior to the disaster, *Donora Death Fog* serves as a cautionary tale about modern-day fealty to corporations at a time when corporations have amassed unprecedented political and social power.

Jennifer Richmond-Bryant, PhD
Associate Professor of the Practice
Department of Forestry and Environmental Resources and the Center for Geospatial Analytics
North Carolina State University
Raleigh, NC

PREFACE

THIS BOOK ORIGINATED FROM AN EVENING SPENT WATCHING TELE-
vision with my wife. We were engrossed in *The Crown,* a fictitious dramatic
series on Netflix that examined the life of Queen Elizabeth II. During the
fourth episode I was struck by a particular conversation—a somber scientist alerting Prime Minister Clement Atlee to the dangers of a smog then
gripping London. The scientist begins the conversation with, "Does the
name Donora mean anything to you?"

I thought, *Not really.* But Atlee knew and went on to explain that it
was a small mill town near Pittsburgh that had suffered an "anticyclone,"
another name for a temperature inversion, an atmospheric condition in
which warmer air traps cooler air below. The "freak anticyclone," as Atlee
put it in the scene, trapped emissions from a zinc factory. "A number of
people died, and several thousand became seriously ill from the poisonous fog."

How had I not heard of Donora before?

I soon discovered that the poisonous fog and the deaths it caused
over Halloween weekend in 1948 had sparked the first clean air act in
the United States, legislation then called the Air Pollution Control Act of
1955. The act was a pivotal event in the nation's struggle to deal with an
environmental crisis. The story of Donora fascinated me then and continues to fascinate me today.

Throughout those six days in October the residents of that valley
town didn't think the smog was a big deal, even though twenty-one of
their neighbors died in it. No one saved hospital records from the time.

Physicians who treated patients in their offices or at the patients' homes either didn't record the visits or destroyed as a matter of routine whatever records they might have kept. And except for two exemplary journalists, Bill Davidson and Berton Roueché, the overwhelming majority of reporters covered the smog as something that happened to a *town*, rather than to its people.

I have tried here to reconstruct the entire disaster as clearly and accurately as possible, using not just the excellent Davidson and Roueché articles but also a wealth of other primary and secondary sources. I have purposely avoided writing a complete history of Donora or any kind of treatise on the environmental impact of the steel and zinc plants there. Readers looking for a more comprehensive history of Donora and its mills should visit the Donora Historical Society, online or in person. Those looking for more in-depth data on the area's ecological history should read environmental epidemiologist Devra Davis's splendid work, *When Smoke Ran like Water: Tales of Environmental Deception and the Battle against Pollution*.

The reader will encounter herein a brief history of Donora during its first few decades, as well as descriptions of the kinds of conditions workers faced in the factories. Only by understanding how Donora came to be and who it drew to its mills can the reader understand why residents reacted as they did during and after the smog: as if the conditions they faced were simply a part of life and not something that could have been prevented.

As for the smog itself, I have striven to present a precise, and I hope complete, breakdown of what happened during those six deadly days, along with a summary of the smog's immediate aftermath. The reader will then find an overview of air pollution today.

As researchers and authors far more competent than I have long known, the spelling of names recorded in official documents before World War II can be notoriously inaccurate. Where discrepancies have arisen I have used spellings I believe are most accurate, or at least the most accepted. Any misspellings of names, then, are my fault entirely and purely unintentional.

There is much more information about the smog, as well as far more stories about Donora's history and Monongahela Valley residents during those deadly days, than I could possibly include here. If I have omitted the reader's or a relative's story, I beg forgiveness.

Finally, I urge readers to contribute in whatever way they can to ensuring that our air becomes ever cleaner, that our land and water become ever less polluted, and that our planet becomes ever more habitable. We owe nothing less to ourselves, our children, and the salvation of our species.

PROLOGUE

HELEN STACK WOKE UP THAT FOGGY MORNING WITH A COUGH AND
sore throat and thought she must be developing a cold. She dressed and
headed down the hill to her work as an office assistant for two of the
town's eight physicians, Ralph Koehler and Edward Roth. The attractive
twenty-eight-year-old arrived before the doctors, as she typically did. The
office looked dirty and was covered in a film of odd dust. "It wasn't just
ordinary soot and grit," she explained. "There was something white and
scummy mixed up in it. I almost hated to touch it, it was so nasty looking.
But it had to be cleaned up, so I got out a cloth and went to work."

With that task finished, Stack sat at her desk to sort through the mail,
then lit a cigarette. It didn't taste right. She took another puff and started
coughing. Hacking. She felt nauseous. "I'll never forget that taste," she
said. "Oh, it was awful! It was sweet and horrible, like something rotten.
It tasted the way the fog smelled, only ten times worse. I got rid of the
cigarette as fast as I could and drank a glass of water, and then I felt better."

The fog was indeed worse than usual, darker than usual, but the next
day's annual Halloween parade was still on. From the office Stack could
hear workers scurrying about, preparing the main drag, McKean Avenue,
for the event. "I knew the committee wouldn't be going ahead with the
parade if they thought anything was wrong," said Stack. "So I went on
with my work, and pretty soon the doctors came in from their early calls,
and it was just like any other morning."

Ralph Koehler often looked out his bathroom window in the morning toward the mills and would watch as plumes of smoke from passing trains drifted into the sky. That morning, though, the morning of Friday, October 29, 1948, Koehler looked out and saw something different and unsettling. That morning, at about 8:30 a.m., a freight train was making its way north, slowly, through the dark, foggy morning. Something about it caught Koehler's eye. "It was the smoke," he said. "They were firing up for the grade and the smoke was belching out, but it didn't rise. I mean it didn't go up at all. It just spilled out over the lip of the stack like a black liquid, like ink or oil, and rolled down to the ground and lay there. My God, it just lay there!"

Koehler was a man of science, and at age forty-eight an experienced one at that. The sight of the train's lethargic smoke provoked his anger. *Well, goddamn,* he thought to himself, *and they talk about needing smoke control up in Pittsburgh!* "I've got a heart condition," he said later, "and I was so disgusted my heart began to act up a little. I had to sit down on the edge of the tub and rest a minute." Koehler could not have imagined what the day had in store for him.

Throughout that Halloween weekend, residents of Donora, a small mill town along the Monongahela River in southwestern Pennsylvania, went about their daily lives as if nothing unusual was happening. The annual Halloween parade went on as scheduled, though people could barely see the floats through the thick, foul-smelling fog. The high school football game took place, as scheduled, between Donora and arch-rival Monongahela, a town that would later boast of having Hall of Fame quarterback Joe Montana among its ranks. Longtime Donora resident Paul Brown recalled, "You could see them punt the ball, hear them kick it, but it would disappear into the cloud."

It wasn't until mortician Rudolph Schwerha received a call at 2:00 a.m. Saturday morning that the weekend began to take on a cloud of despair. The call woke him from sleep, but he was accustomed to it. Schwerha had been a mortician for several years and had just been promoted to chief deputy coroner for the county in February. He did his best to comfort the caller and then sent his driver to pick up the body. "He was gone forever," Schwerha remembered. "The fog that night was impossible. It was a neighborhood case—only two blocks to go, and my driver works quick—but it was thirty minutes by the clock before I heard the service car in the drive." Just as the car pulled in Schwerha received another call. Someone else had died.

Surprised that two people would expire in such a short time, he decided that this time he had better go along with his driver. The second body was in Sunnyside, a community on the other side of the Monongahela River, too far for one man to drive in such weather alone. Traveling would prove not just difficult but dangerous as well. "The fog, when we got down by the mills, was unbelievable," said Schwerha. "Nothing could be seen. It was like a blanket. Our fog lights were useless, and even with the fog spotlight on, the white line in the street was invisible. I began to worry. What if we should bump a parked car? What if we should fall off the road? Finally, I told my driver, 'Stop! I'll take the wheel. You walk in front and show the way.'"

They continued like that for two miles, a distance that might have seemed to them never-ending. Finally they reached Sunnyside, but so thick was the fog that they unknowingly passed the deceased person's house. "I know that section like my hand," Schwerha said. "So we had to turn around and go back. That was an awful time. We were on the side of a hill, with a terrible drop on one side and no fence. I was afraid every minute. But we made it, moving by inches, and pretty soon I found the house." It was the house of a sixty-seven-year-old retired coal miner. Schwerha had no way of knowing then that the miner was the night's third death, not its second.

"When we were ready, we started back," Schwerha recalled. "Then I began to feel sick. The fog was getting me. There was an awful tickle in my throat. I was coughing and ready to vomit. I called to my driver that I had to stop and get out. He was ready to stop too, I guess. Already he had walked four or five miles. But I envied him. He was well, and I was awful sick. I leaned against the car, coughing and gagging, and at last I riffled a few times. Then I was much better. I could drive."

When they finally returned to the funeral home, they found Schwerha's wife, Helen, waiting at the door. "Before she spoke, I knew what she would say. I thought, *Oh, my God—another!* I knew it by her face. And after that came another. Then another. There seemed to be no end. By ten o'clock in the morning, I had nine bodies waiting here." Then he heard that two other funeral homes in town each had a body. *Eleven people dead!* thought Schwerha. *What was happening?* "We didn't know. I thought probably the fog was the reason; it had the smell of poison. But we didn't know."

Donora had been home to steel and zinc plants since the turn of the century, when the town was founded. Toxic chemicals and gases had been spewing out of the factories' smokestacks all day, every day, for decades. Donorans had become so accustomed to smoke and soot that they found the intensely dark fog that Halloween weekend not terribly unusual. "Remember," said longtime resident Harry Loftus, "these guys stormed the beaches of Normandy. Do you think a little smoke is going to bother them?"

The fog, as they called it, had been hanging over Donora and its neighbor, a hamlet across the river called Webster, for four days. It was the result of a temperature inversion, the same weather event that causes morning fog in spring and autumn. This particular temperature inversion, though, was different; it hadn't "blown off," as typical inversions do, nor would it for two more days. In those six days twenty-one area residents perished and thousands were sickened.

The hospital in nearby Charleroi overflowed with patients struggling to breathe and quickly ran out of oxygen tents to treat victims. Firefighters in Donora ran out of oxygen tanks and had to send to neighboring towns for more. Doctors exhausted themselves shuffling almost blindly from house to house to give the sick a shot here, a prescription there. One physician all but succumbed himself before finally reaching home and sleeping a few hours, only to awaken and head out again to treat more patients.

The tragedy was finally announced to the nation by newsman Walter Winchell that Saturday night, the first time many people in Donora, along with their relatives, realized that this fog was different than any that had come before. The deaths that weekend forced the government to finally address a problem it had been struggling with for years—air pollution.

A year and a half after Donora, President Harry S. Truman organized the first-ever national conference on air pollution. The conference led directly to the Air Pollution Control Act of 1955, the precursor to the historic 1970 version. That version, heavily amended after 1955, was signed into law by President Richard M. Nixon and renamed the Clean Air Act, the name we know today.

The Donora smog event was neither the world's first nor its last air pollution disaster. Belgium's industrialized Meuse Valley had suffered a prolonged temperature inversion in 1930—two decades before Donora—that killed sixty people and sickened 6,000. A month and a half after Donora, a six-day smog settled over London and killed 300 people. Another deadly smog in London, this one four years after Donora, lasted five days, killed as many as 12,000, and sickened tens of thousands more. London failed to eliminate the root causes of its smog and suffered another prolonged temperature inversion in early December 1962 that left 60 dead.

Despite the technical advances in pollution control that have been made since then, and in spite of a wealth of clean air legislation both in the United States and abroad, air pollution remains a deadly scourge. Air pollution and the varied health issues associated with it continue to endanger lives, especially among people in China, India, Pakistan, Southern Europe, Russia, and Australia. Particularly severe problem areas in the United States include the southern Mississippi River area, the Pacific Northwest, California, and Alaska.

Donora today is, sadly, a shell of its former self. Its mills have long closed, victims of antiquated equipment and newer, more efficient processes, increased environmental regulations, and reduced demand for the steel and zinc they once produced. Banks, restaurants, pharmacies, and clothing stores no longer dot the main street. The mills lining the banks of the Monongahela have been replaced by a mini industrial park, and as lifelong Donora residents pass on or move away, newer residents move in and commute to work elsewhere.

There remains in residents, however, a palpable pride in being Donorans, citizens of a town that stands, however tragically, as a reflection of past greatness and a herald of hope for even cleaner air to come.

DONORA DEATH FOG

PART I

ORIGINS

1

DONNER TAKES
THE REINS

THE HARDSCRABBLE HILLSIDE TOWN OF DONORA WAS FOUNDED AT THE terminus of America's Gilded Age, a time when a scattering of unimaginably wealthy individuals began coasting on their monetary laurels, soon to become legendary benefactors and philanthropists. They became known as robber barons for the monopolies they created and the legally and ethically questionable tactics they used. They included the likes of J. P. Morgan, Andrew Carnegie, Cornelius Vanderbilt, Andrew Mellon and his even wealthier brother, Richard, and John Davison Rockefeller, the wealthiest of them all, even by today's standards.

The founder of Donora, William Henry Donner, might not have been a robber baron, but he was close. A slight fellow with deep blue eyes, thick eyebrows, and a pleasant smile, Donner had gained a degree of wealth investing in real estate in the 1880s. He had taken full advantage of a boom in the development of natural gas factories in northeast Indiana, purchasing about 150 properties in Jonesboro and the appropriately named Gas City. He had no intention of settling there permanently, not at all.

For the most part Donner's investments proved profitable, most of the land being sold for considerably more than his purchase price. The young businessman began making a name for himself. Donner was coming of age during what would become known as the second period of industrialization, from the late 1800s to the early 1900s, a time marked by laissez-faire capitalism. Industrial leaders conducted business then with little or no governmental oversight. It was a glorious time for free-market capitalists, a time of wealth beyond measure, but that wealth was built mostly on the backs of common laborers, men and women whose yearly earnings wouldn't be enough to fill Andrew Mellon's vest pocket.

In any case, Donner bought low and sold high, and used the profits to subsidize his next venture, the manufacture of tinplates. Tinplates are thin sheets of iron or steel covered with a layer of tin to protect the underlying metal from rusting. In the nineteenth century tinplate was used to produce pots, pans, cans, stoves, candlesticks, kettles, and all sorts of tableware.

There was virtually no tinplate manufactured anywhere in the United States prior to 1890, however, so manufacturers were forced to rely on tinplate imported from Great Britain, Wales, or Germany. Donner believed that tinplate could be a profitable business if a plant could be built near a steady, inexpensive form of fuel. With a large number of productive coal mines throughout western Pennsylvania, Donner aimed to build a tinplate plant in the middle of those mines, in an area along the Monongahela River south of Pittsburgh.

Donner's plans changed, however, during a trip to Europe in 1898. Donner stopped first at a tinplate works in Wales. He wanted to better understand the manufacturing processes used in Britain before he returned home to build his own plant. From Wales he traveled to London, and there happened upon a good friend, John Stevenson Jr., who later became part owner of Sharon Steel Company, a steel factory on the Pennsylvania border near Youngstown, Ohio. Donner recalled that he and Stevenson met at what he remembered as Victoria Hotel, but what was probably the Grosvenor, located above Victoria Station. The two men soon began talking business.

Stevenson had recently built a plant to produce steel rods, wires, and nails in New Castle, Pennsylvania. The plant was considered at the time the largest such mill in the world. He told Donner that if he was going to build another rod, wire, and nail plant, "it would have a continuous

FIG. 1.1. William Donner at about age thirty-four. Courtesy of Donora Historical Society.

roughing mill and a Belgian finishing mill similar to the Joliet plant," the Joliet Iron and Steel Works, the second-largest steel plant in the country. Built in 1869, the Indiana plant manufactured rails for trains and numerous other steel and iron products.

The idea of producing steel rather than tinplate so intrigued Donner that he changed his plan. "After considering my financial position," Donner said, "I decided to build a rod, wire, and nail plant in which I would have a substantial interest. I knew that the American Wire and Nail Company of Anderson [Indiana] and the Consolidated Steel and Wire Company [in Cleveland] were money makers." Donner, along with his mentor and business partner Andrew Mellon, decided to purchase a number of properties in West Columbia, a sparsely populated area along the inside of a horseshoe-shaped curve in the Monongahela River.

———————

Pre-contact Native Americans who inhabited the area, a people known variously as Alligewi or Monongahela, called the river Minaugelo, or "River with High Banks." Locals now refer to the river as simply "the Mon." The Mon meanders northward from West Virginia coal country to the middle of Pittsburgh, the epicenter of the steel industry, where it joins the Allegheny River to form the Ohio, a confluence known as Three Rivers. Of the three rivers, the Mon is the shortest and by far the muddiest.

The Allegheny's riverbed is largely rock, whereas the Mon's bottom is soft and silty, ready to loose its sediment with a good rain. After a strong rainstorm the churning Mon can send chocolate-brown water streaming northward to Pittsburgh, where it meets the blue-gray Allegheny water flowing south. On certain days the Ohio can seem like a split-personality river, steely blue on the west side and a muddy, cloudy, unappealing umber on the east.

The Monongahela River flowing past West Columbia formed over the millennia a steep-walled valley that rises 480 feet above the river on the western side and nearly 560 feet on the eastern side. Both sides of the river are steep, with the eastern side having a considerably sharper grade than the western. It was on the western side of the river's bend, about twenty miles as the crow flies south of Pittsburgh, that Donner and Mellon decided to build a brand-new town, a town created solely to make steel.

The town's name is a merger of William Donner's last name and the first name of Andrew Mellon's wife at the time, Nora Mary Mellon. Donora was then and remains today the only town so named in the world. It sits directly across the river from the tiny community of Webster, a town that shared in Donora's growth as well as in its sad decline.

To create the town, Andrew Mellon set up a trust in early 1899. The word "trust" in those days didn't mean what it does today. Trusts today are used to administer a person's legal and financial affairs, typically after a child reaches a certain age or after a person's death. During the latter half of the 1800s, however, and through the early 1900s, trusts were used to consolidate power among a group of investors. Shareholders in companies owned by the investors would assign their shares to specific trustees, who were then given carte blanche to make business decisions on their behalf. Trusts became a highly regarded tool for avoiding governmental control of corporations.

Mellon's Union Transfer and Trust Company, later renamed the Union Trust Company, could manage businesses under its umbrella as the trustees saw fit, without interference or oversight from the government. If a trust controlled several banks, for instance, it could dictate interest rates for each bank, thereby squelching competition from lesser institutions. Just so did trusts allow monopolies to develop, using sheer, overwhelming market power rather than open and honest competition.

In 1890 the government tried to clamp down on unfair competition and prevent monopolies with the Sherman Antitrust Act. The act was proposed by Ohio senator John Sherman, who served thirty-two years in the US Senate. Antitrust sentiment had been growing throughout the country during the late 1800s, and Sherman had led the charge to reign in those trusts. However, the act failed to define such critical terms as *monopoly*, *conspiracy*, or even *trust*.

A stronger bill in 1914, the Clayton Antitrust Act, would prove significantly more effective in blocking monopolies. That act would prohibit anticompetitive mergers, block predatory pricing, allow individuals to sue companies for illegal or unethical corporate behavior, and promote key labor rights, such as unions' right to organize and to peaceably protest. Until the Clayton Act was passed, though, Mellon could continue to use trusts to shield his interests. Through his Union Trust Company, Mellon set up a separate trust for the Donora project in early 1899. That trust, named the Union Improvement Company, would control virtually everything in the new town, and in charge of the Union Improvement Company, Mellon placed William H. Donner.

Donner immediately assumed responsibility for nearly all aspects of the development of the new town, from the maps of home lots to the roads and streets in town to the homes where mill officials would live, and even the makeup of the town's own government. His decisions would affect the life of every resident for decades to come.

BREAKING
RECORDS

DONORA

Is the name of the new town founded by the Union Improvement Company in the great Monongahela Valley, and is located on the line of the Pittsburg, Virginia & Charleston Railroad, midway between Monongahela City and Charleroi, at which point the

UNION STEEL COMPANY

Are now erecting the largest and most complete wire rod and nail works in the world and in order to properly provide for the vast army of men that will be employed in their works the Union Improvement Company, being an outgrowth of the Union Steel Company, have already contracted for a large Brick School House to cost $30,000, a large business block to be known as the "Postoffice Block," a fine Bank Building, etc., etc., in addition to which they already have the assurance that several very handsome residences will be built at once. The town will be modern in every respect, paved, sewered, and in fact everything done to make it a healthful and desirable place to live.

UNION IMPROVEMENT CO.,
1223 CARNEGIE BUILDING.

JUST SO RAN ADVERTISEMENTS IN PITTSBURGH AND OTHER NEWS-papers in the spring of 1900 in preparation for the first sale of home lots in Donora. Groundbreaking for the rod, wire, and nail mills occurred on May 29 that year. Donner planned to locate all of his mills in Donora, on the western side of the Monongahela. None would be built in Webster, the community across the river. He located three types of furnace at the southern end of the mill complex: blast, Bessemer, and open hearth. The three types of furnace each worked somewhat differently and produced steel in a variety of compositions, depending on a customer's needs.

Donner probably chose the southern end partly because of where most of the coke used by the furnaces would come from. Coke, a form of nearly pure carbon, is critical for making steel. A wide swath of coke deposits runs diagonally from the middle of Alabama north through West Virginia into northern Pennsylvania. Much of Donora's coke came into town from the south, mostly from coal and Pennsylvania coke towns farther south and west than Donora, and perhaps some from West Virginia as well. Much of that coke, as well as limestone and raw iron ore, reached Donora on barges that traveled down the Mon through a series of nine locks and dams. Barges would pull into Donora's docks and unload their wares onto freight cars, which would then distribute the coke, ore, and limestone to furnaces throughout the mill complex.

The furnaces burned coke, ore, and limestone to make molten iron. From there the iron would be sent to a series of steel mills for processing. Donner placed those processing mills—the blooming mill, nail mill, rod mill, and wire works—north of the furnaces. The steel mills produced an assortment of products, from construction nails to sheets of steel to Ellwood fences, a type of fence made of wires woven in a repeating diamond pattern. American Steel & Wire advertised its Ellwood fencing products for anyone "needing an efficient farm, field or ranch fence, secure against outbreaking or inbreaking horses, cattle, hogs, pigs, sheep, dogs, poultry or rabbits."

Donner, wanting to prove his worth to Mellon, expected contractors to "break records" in erecting the furnaces and mills, and they did. While Donner was dispensing orders to mill contractors he was also working to populate the area with workers. At 10:30 on the morning of Thursday, August 30, 1900, a gun was fired and hundreds of potential homeowners flooded the Donora land office to lay claim to their chosen lots. These were the men and women who would run the mills and furnaces, staff the stores and hotels in town, and give birth to the next generation of mill workers.

Inundated by the surge of people at the door, land office clerks must have felt overwhelmed at the prospects of managing all of those claims as quickly as Donner demanded. They handled a total of $225,000 in home sales that day alone, a staggeringly high number for the time and location.

Some lots were sold before the public was invited to bid. Lot 1, between First Street and Highland Avenue, was sold on August 13 to James Castner Heslep, who would become instrumental in the town's early growth. James A. Taylor and Archibald M. McCrea bought lots 11 and 39 the same day, most likely for their homes, and returned four days later to purchase a block of lots, 39 through 42, in the heart of Donora, between Sixth and Seventh Streets. The men spent a total of $4,160 for all six lots. On August 25, the *Monongahela Valley Republican* noted that a John Koocka and John Milka, both of Bentleysville (now just Bentleyville), Pennsylvania, bought two lots, 26 and 34, for a total of $165. The lots fronted Meade Street at the top of the Donora hill, which accounts for the considerably lower price.

Real estate prices in Donora, as everywhere else, were determined largely by location. Here, lots closer to the bottom of the hill—and the mills—were deemed more valuable; those farther away, less so. Several lots were reserved for mill officials, including one for William H. Farrell, the first superintendent of the wire mill, the first mill built in Donora. The homes of each mill superintendent received a prominent lot close to the main thoroughfares, McKean and Meldon Avenues.

Roman Earl Koehler was there when a second batch of home lots went up for sale. "Some would-be purchasers appeared on the scene the night before," Koehler recalled, "and remained beside a lot stake until the opening gun was fired the next morning at ten. Then the lot ticket was removed, and a rush made to the land office to arrange for the purchase. In many instances there were two or more men camped on the same lot, and there were a few battles over possession of the lot ticket."

The lots had been mapped out by John G. Parke Jr., an engineer who had made a national name for himself during the tragic Johnstown Flood in 1889. Parke had been the resident engineer at South Fork Fishing and Hunting Club, a summer retreat for such monied industrialists as Andrew Carnegie, Andrew Mellon, Henry Clay Frick, and James M. Schoonmaker, who had named the town of Monessen. An enormous rainstorm that summer had caused an earthen dam at the club to burst, loosing twenty million tons of water into the valley below and unsuspecting Johnstown. The torrent crushed houses in its path "like eggshells," tossed railroad cars into the air like toys, and splintered huge trees "like pipestems."

More than three thousand people died in the flood, not just from the deluge of water but also from an inferno that broke out among the debris that had crashed against a stone bridge in the center of Johnstown. Many more people might have lost their lives had it not been for twenty-two-year-old Parke, who jumped on his horse and galloped downstream to warn residents below of the unfolding catastrophe. Despite the deaths and indescribable damage to the valley, newspapers throughout the nation found redemptive news in the heroics of the young Parke. The *Boston Globe* called him the "Paul Revere of South Fork."

In planning Donora, Parke planned two main avenues running parallel to the river bend, Meldon and McKean. Meldon was closest to the mills and was named in honor of Andrew Mellon and William Donner. McKean runs a block west and parallel to Meldon and was named after Donner's colleague, James McKean, who had introduced Donner and Andrew Mellon. The north-south avenues intersected with streets that ran east-west, from the floodplain to the top of the hill. Streets were numbered sequentially from south to north, with First Street at the southern end and Fourteenth Street at the northern.

Several of the streets are acutely steep, too steep for travel by car, much less by horse and buggy. Part of Fifth Street, for instance, was closed to vehicular traffic in the mid-1900s because the incline proved too steep for cars and trucks. Pedestrians today navigate the hill by walking up or down the middle of the still-paved street or treading a set of stairs, one on either side of the street. Stairs on one side have been replaced since the town's inception, but stairs on the other side remain original to their installation. That stairway consists of 196 steps with unusually short risers, half the height of modern stairs.

Elmer J. Iiams, an engineer in Donora, designed the stairs and set the height of the risers at precisely four inches to accommodate the day's fashions. Hobble skirts, with their exceedingly tight hems, were all the rage at the time. Making the risers too high would have meant that hobble-skirted ladies might raise their feet too high as they shuffled up the stairs, thereby showing an unseemly amount of forbidden ankle. Iiams's granddaughter, Rosemarie, explained her grandfather's intentions. "His wife was a little woman," she said, "and his daughter was a big woman. So he measured the distance that each could raise her legs in her hobble skirt and made the steps halfway between."

Construction crews building homes in Donora needed creativity to manage the hill's inclines. Many houses were designed with steps between the dining room and kitchen or from one level to another. Houses might have many stairs leading to a side door to the street but have just one or two steps from the front door to the street. Most houses were built extremely close together as well, so that more homes could be built in the geographically limited area of the town.

Stair and street steepness played a substantial role in the everyday life of any Donoran. "Pedestrians had to walk on hundreds of steps to go up or down the hill," recalled former resident Sidney Mishkin. "Handrails were provided to assist the pedestrians. Bypass-like streets were built around Fifth Street hill and other hills so that automobiles could drive around the hills rather than straight up or down, even if [the roads] were open to traffic." Anyone walking those hills every day would have developed strong leg and hip muscles and robust lungs without ever setting foot in a gym.

During the earliest years of Donora homeowners needed guidance about what could be built on their property and where. In a brand-new town, all sorts of decisions needed to be made, including how disputes between neighbors should be resolved, how new stairs or sidewalks should be designed, which paint colors for homes would be allowed, what items in a person's yard could be left in place and for how long. In the early years one of the more important discussions landowners had with town officials concerned where and under what conditions a landowner's cows and sheep could graze. "That was a huge thing," explained Donora historian Brian Charlton. "What kind of fence were you supposed to use? How much area could you use? When could you take your cow out for a walk? Just all kinds of things."

Town officials, most of whom had been appointed either by William Donner himself or by one of his associates, made all those determinations. Donner's men—there were no women in charge—set the precedents for how an enormous variety of local conflicts would be resolved for years to come.

3

A TOWN
BLOSSOMS

FROM AUGUST 1900 THROUGH THE END OF 1902 CONSTRUCTION CREWS
working on houses and mills fairly owned the streets of Donora. Trees
were leveled. Trenches were dug for sewer and water lines. Trucks carrying lumber and building supplies rumbled along dirt roads.

Such a cacophony of sounds there must have been: hammers slamming
nails into two-by-fours, cement mixers rumbling, tractor engines throb-
bing, the cracking of trees falling and the chorus of snapping branches
as the crowns crashed and broke apart, bricks being slapped one by one,
row upon row, to create walls for the mills more than a hundred feet high
and hundreds of feet long. Foremen shrieking decrees. Workmen of all
kinds chatting, cursing, and yelling to one another in a never-ending
stream of work-talk. There were carpenters, plumbers, painters, tilers,
roofers, masons, electricians, surveyors, welders, engineers, ditch diggers,
and millwrights working together to muscle an area that had once held
just four dwellings and a dozen souls into a bustling town of a thousand
homes and six thousand people—all in just three years.

Donora became in the blink of a dust-covered eye one of the most desirable towns to live and work in western Pennsylvania. Houses lined the hillside. Businesses along McKean Avenue burst forth like umbrellas on a Seattle winter's day. "It is impossible for one who has never visited the town," wrote newspaperman Roman Koehler, "to realize the extent of the improvements or even for one who has seen the place to realize that all this has been accomplished in so short a time. There has, probably, never been another instance in the history of the country where a community has attained so rapid, and at the same time, such substantial growth."

Among the first businesses in town was First National Bank of Donora, which built a prominent three-story brick building at the corner of Sixth and McKean. It began business July 15, 1901, with a total capital of $75,000, deposits of $300,000, and $500,000 in total reserves. Members of the board of directors included, unsurprisingly, William Donner and Andrew Mellon, as well as Bert Castner, the head of one of Donora's most influential families.

The Castner clan had been living in the area since before the Revolutionary War. The "original" Castner was Peter, born in 1733 in Bucks County, Pennsylvania, north of Philadelphia. He married Mary Magdalena Rugh and moved first to Latrobe, Pennsylvania, and from there to Walnut Bottom, forerunner to West Columbia. Peter and Mary staked their claim to about 140 acres of land, where they built a home and farmed the land. They had five children together: Maria Catherine, John, Elizabeth, Mary Magdalene, and Jemima. Each of those children produced Castners of their own, and so on until, by the time Donora came into being, walking down the street without bumping into a Castner would have proven rather difficult. In fact, the first child born in Donora was Rebecca Donora Castner, who arrived July 23, 1900, and would one day figure prominently in a huge celebration in town.

Most eminent of the Castner brood in those early days was Peter and Mary's great-grandson, Bertrand Wilson Castner, or Bert, who served as the town's first burgess and who, with son James, little Rebecca's father, owned an insurance and real estate firm. Bert served as director of the Union Improvement Company, vice president of the First National Bank of Donora, and justice of the peace. He had inherited his family's acreage in the late 1800s, and in 1900 sold most of it to the Union Improvement Company for the new mills.

Bert and James's real estate company faced competition early on, not that it could have handled all of the town's real estate needs by itself anyway. The firm of Schooley and Kenyon entered the "constantly widening circle of Donora's business contingent" in July 1901. John R. Schooley and J. C. Kenyon acted on behalf of numerous out-of-town landowners. They managed rentals and repairs for their clients' existing homes, monitored construction of new homes, and mediated relevant tax issues. Like the Castners, the Schooley firm sold not only homes but also insurance, a common practice of the day.

Another combination business sold furniture alongside, of all things, caskets. E. G. Sturgis operated Sturgis Furniture and Undertaking at the corner of McKean Avenue and Sixth Street. The basement and first floor were "literally crowded with the best grades and most finely furnished product of the cabinet-maker's art." Sturgis reserved one part of the store for a collection of caskets of "the most elegant and artistic designs obtainable in oak, mahogany, etc., in various styles of upholstery, forming a selection from which any taste or purse may be suited." A morgue was conveniently situated at the rear of the building and had, thankfully, its own entrance. The business catered to Slovaks in the area. A sign at the front of the building read, in slightly misspelled Slovakian, SLOVENSKY POHRABNIK, or "Slovak undertaker."

One of the first and most successful businesses in town was the Donora Lumber Company, founded by Charles Potter of nearby Charleroi and several businessmen from Pittsburgh. Wood from the company was used to build many houses in town, the building that would house the First National Bank, later Mellon Bank, and the town's community center, a building that would perform a quite different function during the 1948 smog.

The first clothing store was The Famous, owned and operated by William Altman. The store opened March 1, 1901, and carried such items as trousers by Cleveland and Whitehall and overalls by a company daringly named Kant Rip.

A trolley line running along the center of McKean Avenue from First Street to Fifth Street connected all of the businesses in town. The trolley car serving Donora, known as "Old Maude," began operating December 15, 1901, and ran on tracks set along a kind of yellow brick road. In Donora's early days highways and streets were made with bricks rather than cobblestones, bricks being more plentiful and less expensive. The color of the bricks varied depending on what was available. Brick refractories in the area produced yellow, red, and orange clay bricks, which were then sold to contractors throughout the region. Yellow bricks were used to top McKean Avenue, over which Old Maude rumbled.

Donora's trolley line was a single spur originating in Donora and linking, via an extension completed in 1911, to the Pittsburgh Railways interurban system a few miles north of town. During the summer, interurban trolleys would stop at Eldora Park, an amusement park located at the junction between the Donora line and the interurban system. The park proved highly popular among members of the growing working class in the area.

The early 1900s saw an explosion of growth in industries and urban populations. "One of the many benefits . . . of urbanization and industrialization," explained historian Brian Charlton, "was the development of leisure time for the laboring masses. What was once the exclusive domain of the 'Idle Classes,' wage laborers, working timed shifts, found themselves with two things they never had before: free time and discretionary income." Wage laborers from Pittsburgh and the whole upper Monongahela Valley enjoyed trekking to Eldora Park for some weekend fun.

Five thousand visitors descended on Eldora Park when it opened on May 30, 1904. They rode a genuine Ingersoll wooden roller coaster, built by Fred Ingersoll, who would go on to build forty-four amusement parks and design or build a total of 277 roller coasters nationwide. The coaster at Eldora was a single figure-eight loop of coaster, and visitors lined up to ride it. At the carousel parents stood with their children as they rode brightly painted wooden horses round and round. Families picnicked on the lawn, couples danced in a pavilion that doubled as a roller skating rink in colder weather, and visitors and staff alike were mesmerized by entertainers and world-class speakers, including union organizer Mary Harris "Mother" Jones and Booker T. Washington, founder of Tuskegee Normal and Industrial Institute, now Tuskegee University.

Donora experienced a wealth of firsts in its infancy, from its first plumber, drayman, and grocer to its first bakery, jewelry store, and meat market. Many new Donorans needed to stay in a hotel until their homes were finished being built. Hotel Irondale, the first licensed hotel in town, had forty "well-furnished rooms." Its large bar quickly became a fashionable meeting spot for Donora's younger men. Located at the corner of Sixth and McKean, the hotel served Donora initially as a hotel and later as a rooming house for mill workers.

R. Mitchell Steen Jr., historian and former managing editor of the *Valley Independent*, Monessen's local newspaper, recalled that the Irondale was built "at a glorious time in the town's history, when the new mills were thriving and newspaper advertisements were clamoring for workers and pleading for people to build homes. It was the time of the 'free lunches' offered by all the bars, then an integral part of any hotel. For a nickel beer, the patron could load up on free food, which consisted of lunch meats and cheeses."

Children in the area initially went to the Gilmore School, built after the Civil War on land donated by John Gilmore. Gilmore was a successful steamboat captain and coal mine owner who had served in the war and later bequeathed to the town land for both a cemetery and a school. Gilmore School was big enough for the children living in the area at the time, but it would be overwhelmed when the new town officially opened. So a new school was built on Allen Street, funded by a loan from Donner's Union Improvement Company.

The foundation for the new school was laid before the first home lot in Donora was sold. Soon the Allen School, a brick building with eight classrooms, welcomed its first students. Just over 250 students enrolled in 1901, with 625 enrolling the following year. By 1903 another school was built, the Castner School, a modern two-year high school at Tenth and McKean that would graduate two girls in its first class, Edna Lewis and Nannie Hodge, both of whom would later become teachers.

A variety of houses of worship sprang up to cover the many religions practiced by Donorans. Although there were small churches in the area prior to 1900, the first official place of worship in the new town was the Quinn Chapel African Methodist Episcopal, organized in 1901. Ohab Shalom became the first Jewish congregation in town, organized in 1903. Orthodox congregants wouldn't receive a permanent synagogue until 1911, when a modest brick structure on Second Street was completed.

Saint Charles Borromeo Church was established in July 1902 and had as its first pastor Charles J. Steppling. Steppling had been born in, fittingly, the German steelmaking city of Essen, the same city that helped give Donora's neighbor to the south, Monessen, its name. The town's Polish residents considered Holy Name of the Blessed Virgin Mary, established just a month later, the "symbol of their faith." The building occupied a prominent space on Second Street, its Gothic spires reaching high above the mills.

By 1920 there was in little Donora a synagogue and fully sixteen churches, including ones for Methodist, Presbyterian, Baptist, Lutheran, Jesus Christ of Latter-day Saints, Russian Orthodox, Greek Catholic, and Roman Catholic congregants. It was a place of intense cultural, religious, and ethnic diversity.

In early Donora, as in thousands of cities and towns across the country, a new kind of club emerged that focused on a sense of community, belonging, and support for local interests. Some clubs were based on religion, others on ethnic similarities. The strong ethnic diversity in Donora led to the development of a variety of clubs where patrons could gather for special events, administrative meetings, dances, dinners, and drinks. Especially drinks.

Russians who found their way to Donora formed the Rusin Club at the corner of Fifth and Meldon, a block down from the Sixth Street gated entrance, one of only four entryways into and out of all of the mills. The other gated entrances were located at First, Eighth, and Fourteenth streets. Clubs, bars, and restaurants sprang up as close to one of those gates as possible, the better to attract patrons leaving the mills after a long day's labor. The location of Hotel Irondale at Sixth and McKean, for instance, was no accident. The Sixth Street gate served as the main entrance and exit for four huge mill complexes, including the blooming, nail, rod, and wire mills. Hundreds of workers poured out of that gate hungry and thirsty at the end of the day, and the hotel bar filled with patrons promptly after the mill whistle blew.

One of the more infamous drinking spots in town was at the base of Tenth Street, a short walk from the Eighth Street gate. Its official name was Liberty Cafe, but it was known by locals as the "Bucket of Blood" for the frequent drunken brawls that occurred there.

Immigrants from Slovenia formed the Slovak Club, incorporated in 1915. The Slovak Club was located in a building on Meldon that had a large auditorium, complete with a stage and balcony. The auditorium was a favorite venue for parties, weddings, roller skating events, basketball, plays, dances, and even magic lantern shows. Immigrants from Croatia formed the Croatian Club. Scottish men formed Clan Grant, and Scottish women formed the Broomie Knowe Lodge, the local lodge of the Daughters of Scotia, a group initially formed in New Haven, Connecticut, in 1895. Other clubs included Sons of Italy, the German Political and Beneficial Union, Spanish Club (its pool open to members only, as long as they weren't Black), and Saint Dominic's Men's Club, a club for Roman Catholic men. Many members of those clubs also belonged to the Donora Club, which occupied the third floor of First National Bank at Sixth and McKean.

The Donora branch of the Polish Falcons of America, called Nest 247, was built on the philosophy of *mens sana in corpore sano*, Latin for "a sound mind in a sound body." The Falcons gained notoriety later in the century as having honed the athletic skills of one Stanislaw Franciszek Musial. Stan "The Man" Musial would go on to become one of the greatest baseball players in the history of the sport, winning seven batting titles and earning entry into baseball's Hall of Fame on the first ballot, a remarkable achievement in any era.

———

Ethnically centered neighborhoods didn't form in Donora like they did in New York City and Pittsburgh. Those cities tended to develop into heavily ethnic neighborhoods, such as Chinatown (Asians), Harlem (Blacks), and Little Italy (Italians) in New York and Polish Hill (Polish), Homewood-Brushton (Blacks), and East Liberty (Italians) in Pittsburgh.

There was no Spanish neighborhood in Donora. Nor was there an Italian, Russian, Croat, German, Hungarian, or Slavic neighborhood. From the beginning, when home lots were sold on a first-come, first-served basis, people lived wherever they could find a suitable place to call home. "Although the Spanish did settle near the zinc plant [built in 1915]," explained Charlton, "it was not an exclusive neighborhood. In that same general area of north Donora you could find Poles [such as Stan Musial's family], African Americans [such as the families of baseball legends Ken Griffey and his son, Ken Jr.], Welsh, Scots, Italians, Carpatho-Rusyns, Germans. Everyone lived everywhere." The only ethnically exclusive neighborhood in Donora was Cement City, where only Scottish and

Irish factory foremen and their families could live. "Because of the housing shortage it seems no single group could dominate an area," Charlton explained.

Donora's diversity was perhaps no more evident than on the town's main streets. Everyone walked along McKean, Thompson, and Meldon, their voices forming a symphony of multinational languages. The sounds of people speaking Spanish, Polish, Italian, Russian, German, or one of the Baltic languages mixed mellifluously with the thrum of daily life. "When older adults spoke to one another on the streets of Donora," recalled lifelong resident Sidney Mishkin, "they generally spoke in their native language. Walking in downtown Donora was like walking in a multinational market."

4

MAVERICKS NOT ALLOWED

DONORANS HAVE ALWAYS FELT GREAT PRIDE IN HAVING GROWN UP IN neighborhoods of great diversity, yet the town was not without its ethnic or racial problems. The kinds of ethnic and racial tensions that occurred throughout the nation also occurred in Donora. For instance, fueled by a decades-long wave of Jewish immigrants in the latter half of the 1800s, antisemitism nationwide crescendoed between World War I and World War II. Historians Jonathan Sarna and Jonathan Golden of Brandeis University described how antisemitism typically expressed itself: "Private schools, camps, colleges, resorts, and places of employment all imposed restrictions and quotas against Jews, often quite blatantly. Leading Americans, including Henry Ford and the widely listened-to radio priest, Father Charles Coughlin, engaged in public attacks upon Jews, impugning their character and patriotism."

Antisemitism in Donora occurred as well, perhaps more surreptitiously but hurtful nonetheless. Resident Sidney Mishkin recalled that Donora stores owned by Jews rarely, if ever, used the owner's last name if it sounded at all Jewish. "In a way it was a charade," recalled Mishkin.

"Donora was small enough that everyone knew who owned what, but I suppose Jewish business people generally felt that the more generic the names of their businesses were, the better. Why remind a customer that he had a choice between a Jewish-owned store and one that wasn't?"

Xenophobia in all its forms has been common in the United States since its inception. Even the great Benjamin Franklin predicted that "swarthy" German immigrants would "overwhelm" the nation. No ethnicity or race, however, was demonized more during the first half of the twentieth century than Blacks.

From the time Donora was established to the end of 1948, the year of the smog, nearly 1,800 Black people were lynched. Most of the lynchings occurred in the South, but a significant number occurred in the North, including fourteen in Pennsylvania, forty-seven in Indiana, thirty-four in Illinois, and twenty-nine in Maryland. Donora factories employed many Blacks who had moved north during the Great Migration, a huge exodus of southern Blacks to the North from the early 1900s to 1970. Blacks, particularly those in the rural South, wanted to escape draconian Jim Crow laws. Some of them ended up in the Mon Valley.

Blacks who moved to Donora early in the century had faced racism before and would have probably taught their children, who might never have lived in the South, how to deal with this new world of Whites. Reggie Walton was one of those offspring. Walton, senior US district court judge for the District of Columbia, grew up in Donora and understood his limits from an early age. Walton's father, Howard Walton, had moved from his boyhood home in Rice, Virginia, to Donora for the "economic opportunity" of working menial jobs at American Steel & Wire. Two of Howard's sons, Johny and Preston, also worked there. The elder Walton and his wife, Sadie, were intelligent and hard-working people, but neither could advance past the lowest-paid positions in town. "My mother was unable to get a job selling goods at the five-and-dime store," recalled Judge Walton. "Blacks couldn't get those jobs. No African American could hold a supervisory position in the steel mills." Nor could they dine at indoor restaurants. "I never went to a sit-down restaurant until I went to college," recalled Walton.

Helen Jenkins also grew up in Donora and said, "Throughout my whole life in Donora there were certain norms that we were taught, and there were certain places that we were not welcome to come into. I can remember in 1954, somewhere around there, was the first time I was able to go into a drug store and sit at the counter."

Before the earliest days of the civil rights era, Donora was in some ways similar to towns elsewhere in the country. "We had separate Girl Scouts and Brownies," said Jenkins. "We could not go with the White troops."

Movie theaters discriminated as well. "In one theater in town Blacks were allowed to sit in what would be the orchestra level," said Jenkins, "but they had to sit on the side. There was another theater that had a balcony. That's where Black people had to go, to the balcony. You couldn't sit in the orchestra."

When Marvin Preston attended Donora High School in the early 1950s, he was one of only eight or ten Black students in a class of 180. "I think the kids got along just fine," Preston said. "I think everybody did. I don't recall any problems in Donora when I was growing up, any kind of racial problems. We had everything that the Whites had. There was no need for me to worry about going to something that they had."

Yet Preston knew all the places where he wasn't allowed. He had realized early in his life that White society restricted where he could go, what he could do, and with whom he could be friends. His Black friends knew the limits as well, as did any Black family in Donora, from the town's earliest days through the 1960s and beyond.

Despite those social restrictions, Black residents found ways to enjoy small-town life. "We had a skating rink up in Belle Vernon called the 'Piggy Wiggy,'" Preston said about Riverview Skating Rink, the official name. "Everybody used to go to the Piggy Wiggy. One interesting thing about that was, there was only one night we could go, that Blacks could go. We never paid that any attention, because we had skating rinks all around the valley. Monday night was in Pittsburgh. Tuesday night was a certain city, I forget now. Each night we could go to a skating rink, so we didn't really care."

Born in Pittsburgh in 1936 and growing up in Donora, Preston grasped early on how life for Black kids differed from life for White kids. Preston remembered the time he asked a White girl to dance. At the time all students were invited to school dances, no matter their race, creed, or nationality. That's just how Donora was. "Everybody was there," said Preston. "But Blacks, you know, we didn't dance together [with Whites]. That was not accepted." He broke that unspoken rule once. Just once.

"There was a White girl," he reminisced. "I think we were sort of sweet on each other, I don't know. But anyways, I did ask her to dance, and we just danced. The next day everybody had a heart attack. I was called into the office. My mother told me I had lost my mind. It caused a lot of confusion, believe me." Other than being called to the principal's office and being told by his mother that he was "crazy," Preston faced no other substantial consequences he could recall. The girl he danced with, however, was banned by her parents from any after-school activity. She had to go directly to school and then had to return directly home. Because she had danced a single dance with a Black boy. Even into his eighties Preston felt guilty. "I did dance with that girl," he said, more than a trace of regret trickling through his aging voice, "which I shouldn't have done. I was just a maverick, that's all."

5

BUILDING
THE MILLS

DONORA'S MILLS, AS DONNER HAD DEMANDED, HAD GONE UP QUICKLY. The first mills built were the wire and rod mills, completed in the spring of 1900. Those mills produced steel wires and cables for use in bridges, roadways, and buildings. The wire mill consisted of a wire drawing department, wire nail department, wire galvanizing department, and a varnished wire department. Two steel rod mills were built to create bars and solid and hollow rods.

The wire mill became profitable quickly; it produced wire used in the production of barbed wire, which had proven enormously valuable throughout the late 1800s and, as far as William Donner was concerned, would continue to prove so into the future. Used on nearly every farm and ranch, barbed wire kept livestock inside holding pens, predators away from livestock, and free-ranging herbivores away from wheat fields, cornfields, and other revenue-producing areas. In all, the wire and rod mills produced more than ten thousand separate products.

Donner appointed as the first superintendent of the mill William H.

FIG. 5.1. Interior of a spooling department in a wire mill. Courtesy of Donora Historical Society.

Farrell, an experienced steelmaker from Ohio. Farrell was a distinguished-looking gentleman with a thick, gray, neatly trimmed mustache and dark, piercing eyes. He would go on to become president of the Bank of Donora in 1906. Farrell and Donner decided to purchase a number of Bates wire-making machines, considered at the time the best barbed-wire machines around.

The machine and its patents, however, were owned by American Steel & Wire, at the time one of US Steel's many subsidiaries. In one of several twists of irony in Donora's development, American Steel & Wire would soon become one of the Union Improvement Company's most important partners. For now, though, Farrell and Donner would have to content themselves with a different kind of machine, slower than a Bates but newer in design.

The team bought eight machines at first. Then, when they began producing high-quality barbed wire, the team purchased another fourteen. "We soon received a surprise," Donner wrote in his autobiography. "The

American Steel & Wire Company sued us [Union Improvement Company] for infringing on one of their barbed wire patents." So Donner, Farrell, and an innovative draftsman named Mont Hughes reviewed all of the US patents for barbed-wire machines and discovered that Donner's machines did in fact infringe on the American Steel & Wire patent. Donner pushed his team to find out whether their machines could be retooled to avoid patent issues while still producing high-quality wire. Hughes soon reported he had found a fix that would not only avoid infringement but also double the machine's speed and capacity. Within a few weeks all of Donner's wire-making machines had been retooled and were spooling out barbed wire at breakneck speeds.

Donner, never one to shirk from a bit of competitive fun, visited the Duquesne Club in Pittsburgh and there met with W. P. Palmer, president of American Steel & Wire, to thank him for "getting after us."

"What do you mean?" asked Palmer. Donner told him he was referring to American Steel & Wire's suit against him. "Oh," said Palmer, "Mr. Bennett handles those matters." Donner then explained that the suit had prompted his company to change their barbed-wire machines, "with the result that our capacity has increased fifty percent." He suggested that Bennett send a representative to Donora so that Farrell could show off the machines. Soon thereafter American Steel & Wire dropped its patent infringement suit, and Donner's machines continued to run day and night.

After the rod and wire mills were built, construction turned to the nail, blooming, and fence-binding mills, finished in 1902. Those mills produced all sorts of nails, from three-and-a-half-inch penny nails to dog spikes, or rail spikes, which affix railroad tracks to wooden ties.

While Donner was building the wire mills, a company called Mathew Woven Wire Fence Company was building a factory in town to make woven wire fencing. American Steel & Wire took over the company in 1903 and folded its operations into Donner's wire mills.

Donner constructed factories to manufacture a variety of types of bars, including blooms and billets. Blooms and their smaller cousins, billets, can be rolled into rails, pipes, I-beams, and assorted other construction materials. Throughout the first half of the century American Steel &

Wire's steel bars were used in concrete reinforcement as well as in the manufacture of tools, machines, and equipment. Most of the bars, though, were destined to be turned into parts for cars and trucks.

Donner's mills also produced steel slabs: large, flat, rectangular blocks that can be further processed into plates, sheets, and strips. Plates, the largest steel product made in Donora, were used to build massive construction equipment, skyscrapers, and oceangoing ships. Sheets and their narrower cousins, strips, were produced in long rolls of steel in various thicknesses, depending on their final use. Sheets and strips were used to manufacture cars, trucks, and all sorts of home appliances, as well as office desks, filing cabinets, metal tables and chairs, garbage pails, kitchen utensils, railroad coaches, trolleys, garden hoes and spades, gutters, and corrugated roofing.

Construction of two blast furnaces began in 1902. A blast furnace is essentially a container for chemical reactions. Shaped somewhat like an enormous Erlenmeyer flask, with a wide base and narrow top, a blast furnace allows heated air under pressure to strike a mixture of elements, which sparks a chemical reaction in that mixture and creates certain liquids and releases certain gases. In the case of an iron blast furnace, heated air is forced into the base of the "flask," while various solid elements are fed into the top. Those elements include iron ore, limestone, and coke, the three main ingredients needed to make steel.

Iron ore is mined from sedimentary rock formed nearly two million years ago, when oceans covered most of earth. Ocean water was rich in dissolved iron but lacked oxygen, and ancient bacteria eventually formed that could produce oxygen from sunlight. The oxygen immediately combined with the iron to form the minerals hematite and magnetite, which fell to the ocean floor in alternating bands of iron and silica or shale. These banded iron formations supplied enormous amounts of iron ore. For instance, the floor of Lake Superior and the land of Michigan's Upper Peninsula have been mined for their iron ore since the mid-1800s. Iron ore mined from Lake Erie, about a hundred miles from Pittsburgh, supplied iron ore for many blast furnaces in the Pittsburgh area. Blast furnaces remain to this day a critical component of steelmaking and continue to provide steel for the nation's military and industrial complex.

The second key element in steelmaking is limestone, a whitish or gray

FIG. 5.2. Workers at the nail mill in Donora. Courtesy of Donora Historical Society.

rock composed chiefly of calcium carbonate, formed from the detritus of coral, algae, and shellfish that die in warm, shallow waters. Limestone is used throughout the world for a variety of reasons, including agricultural lime, industrial cement, and such medicines as antacids and calcium supplements. In a blast furnace limestone is added as a flux, an element that removes impurities from the iron ore through chemical reactions inside the furnace. Those reactions depend on the third element, coke. When coal is burned in an enclosed space it releases numerous gases and leaves behind coke, a powerful fuel of nearly pure carbon. Until about 1908 coke was created in beehive ovens, cavelike ovens drilled into the side of a hill. Beehive ovens built into a hillside in Webster supplied coke to Donora during the earliest years of the mills.

Those three elements—combined as roughly four parts iron ore, two parts coke, and one part limestone—are poured into the top of a blast furnace and hit with exceedingly hot air pumped into the furnace from below. When the hot air hits the coke and limestone, the coke burns in the air and heats the iron ore to temperatures greater than 2,800 degrees Fahrenheit. At that temperature the iron ore breaks down into mostly pure iron, called pig iron, and the impurities that had been present in the iron ore. Both the pig iron and the impurities, called slag, drip to the bottom of the blast furnace. Slag is lighter than pig iron and so floats on top of the pig iron.

———————

Old-style blast furnaces were basically large chimneys with an opening to push heated air through and a basin at the bottom. The basin had two outlets, one near the top of the basin and one near the bottom. Every now and then, usually twice a day, a worker would "tap" the outlets, breaking the seal and allowing molten slag to pour out the top outlet and molten pig iron to flow out of the bottom. The slag would be allowed to cool, hauled away from the furnace, and dumped on the ground. The molten pig iron would flow through a series of channels crafted from sand or other suitable material. The pig iron would be allowed to cool in the channels and then broken up into rounded, rectangular blocks somewhat resembling the top half of a pig. The blocks took on the name "pigs," hence "pig iron."

The blast furnaces in Donora were considerably more advanced, though they employed the same essential techniques. Every five to seven hours a worker would tap the furnace with a cast-iron poker. Liquid iron would then flow into one concrete channel, while slag flowed into another. The iron and slag then ran along those channels into separate railroad cars specially manufactured to handle molten material. Shaped like chubby torpedoes, these cars, called bottle or hot-metal cars, were lined with heat-resistant brick. Bottle cars carrying molten iron were pulled by train to another part of the mill, where the iron was processed into steel. Slag cars were pulled to a slag dump, an area generally off-property and well away from residential areas, then unceremoniously dumped. The slag would cool and form gray-black, pockmarked rocks, like lumps of coal with acne.

Construction of twelve open hearth furnaces started at the same time as the blast furnaces. Open hearth furnaces heated pig iron from the blast furnaces to remove impurities remaining in the pig iron. To obtain the high temperatures needed for that process, burners on either side of the open hearth pushed extremely hot air and gas over pig iron in the middle. Waste gases were sent into large chambers, heated, then pushed back through the hearth, adding even more heat to the iron. That intense heat produced steel with a silvery-white sheen. The steel would be poured into molds and sent to the blooming mill or one of the other mills for further processing.

Whether steel from Donora found its way into vehicles made by Ford or Chevrolet, refrigerators from Westinghouse or General Electric, or tractors from John Deere or International Harvester has never been determined. The likeliest scenario is that steel made by American Steel & Wire could have been sent to any of US Steel's thousands of customers.

Groundbreaking for the final Donora plant, the Donora Zinc Works, a direct subsidiary of US Steel and not associated legally with American Steel & Wire, took place June 22, 1915, on land formerly owned by Mary Ann Ammon. Ammon had lived in the woodlands of West Columbia her entire life and, with her late husband, Jacob, had borne a dozen children.

She, Jacob, and their children worked the farm at the northern bend of the Monongahela horseshoe. Jacob plowed the land and planted crops. Mary maintained a vegetable garden and kept house, her children undoubtedly helping wherever they could, with older children working

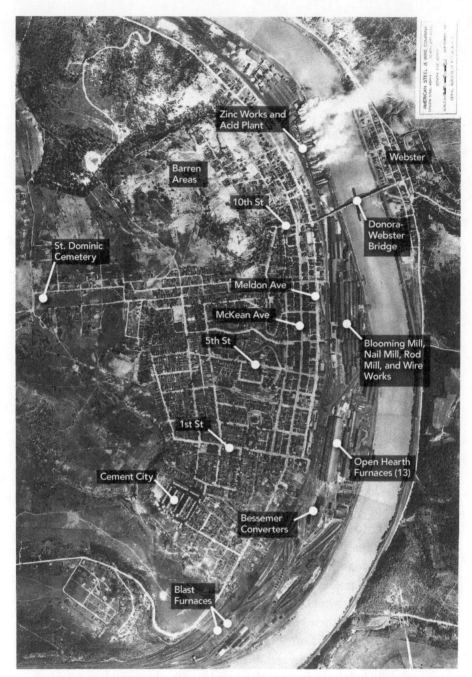

FIG. 5.3. This aerial photo, taken in 1941, shows the location of each main mill in the American Steel & Wire complex. Courtesy of Donora Historical Society.

in the fields and younger ones doing chores under Mary's watchful eyes. Year after year, decade after decade, Mary and Jacob managed their farm and provided for their children.

Jacob eventually became senile, probably sometime around 1905, leaving Mary to assume more and more of his duties. Eventually she had to dress Jacob, feed him, and bathe him, until he succumbed at home to his decaying brain and body at age seventy-eight, on July 19, 1910. Mary must have realized that she could not maintain the farm and homestead by herself, so she sold the farm to US Steel. Mary was the last West Columbia landowner to sell her land for the mills. The land was eventually used to house a complex of buildings for processing sulfuric acid, a byproduct of zinc production.

Ammon spent her remaining years in Donora and died of a stroke at six o'clock in the evening on March 14, 1925. History has not recorded whether Ammon ever gazed upon the Zinc Works with disgust, for having turned her beloved farm into an unsightly mass of brick and cinder-block buildings, or with gratitude, for supplying her and her family with enough money to live with some degree of comfort. Perhaps she would have felt a bit of both.

The Zinc Works produced its first zinc bar a scant 120 days after groundbreaking, on October 20, 1915. War had been raging in Europe since August 1914, shortly after a radical nationalist from Serbia, Gavrilo Princip, shot and killed Archduke Franz Ferdinand and his wife. Ferdinand had been heir to the Austro-Hungarian Empire, a large section of central Europe that had been home to Germans, Hungarians, Czechs, Poles, Slovaks, Croats, and people of many other ethnicities. Tension had been brewing for years before Ferdinand was struck down, and his death ignited a firestorm.

President Woodrow Wilson, in office just over a year when the archduke was killed, wanted the United States to maintain its neutrality. He believed that if it became necessary to enter the war, troops from the National Guard would prove sufficient. Although he wasn't alone in that belief, a great many others wanted the nation to do much more to prepare for possible war. Among the pro-preparations faction were several high-profile figures, including New York City mayor John Purroy Mitchel, general manager of the *New York Times* Julius Ochs Adler, and two sons of former president Theodore Roosevelt, Quentin and Theodore Jr. They believed that war with Germany was inevitable and that the nation should be prepared.

Industrialists throughout America were monitoring the goings-on in Europe as well. They realized that if war ensued, as it surely would, the nation would need to quickly manufacture all kinds of wartime matériel, in particular steel and zinc. Steel is used for an enormous variety of wartime products, from helmets to tanks to aircraft, bombs, and bullets. All of those steel products wouldn't last long in a battle environment without zinc, arguably the second most important industrial product for wartime.

Bluish-white in its natural state, zinc occurs naturally as zinc ore and can be found in abundance in earth's crust, particularly in areas of Canada, Russia, Australia, Peru, and the United States. The most naturally abundant zinc ore and the form most often used for zinc production is sphalerite. Older miners might remember calling the ore zinc blende, blackjack, steeljack, or rosin jack. When heated to about 1,300 degrees Fahrenheit, sphalerite, or its weathered cousin, smithsonite, breaks down into zinc carbonate, a salt that mixes easily with other metals.

Zinc has found numerous uses throughout history. As far back as 3000 BCE Babylonians merged zinc with copper to produce brass, a metal that can be molded easily and transformed into a wide variety of items. Ancient Romans used brass to make coins, pipelines, statues, and decorative items, among many other products. Persians around the turn of the fourteenth century used a solution of zinc to treat eye inflammation. Zinc is used today to treat a wide variety of ailments, including zinc deficiency, ear infections, cold, flu, bladder infections, tinnitus (ringing in the ears), malaria, cataracts, night blindness, diabetes, high blood pressure, anorexia, dementia, bowel inflammation, and a host of other conditions.

Zinc has found myriad uses in industry as well, from paints, plastics, and printing inks to cosmetics, fluorescent light bulbs, and dry cell batteries. The most common use, though, and the use that generated the most interest leading up to World War I, as well as the greatest revenue for the zinc mill in Donora, was for galvanization, a process used to prevent the formation of rust. When steel is exposed to air, the iron in it combines with oxygen to form iron oxide, or rust. Galvanization blocks that reaction.

In hot-dip galvanization, a common galvanizing process, an iron-based product such as a sheet of steel is dipped into and out of a bath of liquid zinc, leaving a thin film of zinc bound to the steel. When the steel is exposed to air, oxygen binds to the zinc rather than to the iron. That binding creates a layer of zinc oxide, an effective deterrent to the formation of iron oxide.

Hot-dip galvanization had been in use since a French chemist named P. J. Melouin presented a paper to the French Royal Academy in 1742 that outlined a process of dipping steel into molten zinc to prevent rust. Melouin said at the time, "I believe this work will be considered useful and will shed a new light on the properties and uses of zinc and tin."

It did indeed shed light on uses for zinc, including its use on seagoing vessels, where rusted nails and joints could have disastrous consequences. For instance, sheets of copper sheathing covered the hull of the last known slave ship, the *Clotilda*. The sheets were secured with galvanized iron fasteners. If the copper sheathing had been secured with nongalvanized fasteners, the iron would have corroded and could have led to a fatal hull failure. To ensure no corrosion occurred, the fasteners were galvanized and countersunk into the hull's wooden frame. The remaining depressions were sealed with pitch and wooden plugs to help prevent water from seeping through the hull.

Galvanization became an immensely important process throughout the first half of the twentieth century, even becoming part of the common vernacular. In the early 1900s military trash cans and water buckets were stamped with the letters G and I, an indication of the material they were made from, galvanized iron. The stamped G.I. is said to have assumed an exaggerated meaning during World War I and beyond, when it came to mean anything government issued. The process of galvanization found more important uses in the military than buckets, of course, including the manufacture of trench-digging shovels, rifles, cannon, ships, and a plethora of other items.

When US Steel finished construction of its zinc plant it immediately became the largest zinc smelter in the world, producing zinc to galvanize iron and steel products of all kinds, items that found their way not only into war in Europe but also into homes, businesses, and military installations throughout the United States.

Donora Zinc Works consumed a forty-acre footprint along the Monongahela, and on it were fifty-nine separate buildings, including a warehouse (building 1), boiler house (30), six roasting furnaces (17–22) for breaking down zinc ore, and three acid chambers (11–13) for processing sulfuric acid. All of those buildings that housed machines to manufacture all of those products needed all kinds of people to run them.

6

PEOPLING
THE MILLS

ROCCA PIA SITS NEATLY INSIDE A VALLEY IN THE MOUNTAINOUS ABRU-
zzo region of Italy, due east of Rome. Stone and stucco houses pack the
town, which at its widest point is barely seven streets across. It sits alone,
its nearest neighbor, the hamlet of Pettorano sul Gizio, three miles to the
north. In the early 1900s Rocca Pia was home to 1,200 people and was
small enough and isolated enough for residents to know, or at least know
of, nearly all other residents. It was here in this quiet, snug, family-centric
village that Bernardo Di Sanza was born.

Di Sanza was the oldest of six children born to Fortunato and Pasqua
Di Sanza. He spent his first twenty-four years in Rocca Pia before heading to the New World. He set sail in 1905 from Naples on the passenger
ship *Roma*, destined for Ellis Island. How, when, and where he met his
wife, Liberata, is lost to history. Like Bernardo she was born and raised in
Rocca Pia. They might have known each other growing up and perhaps
even fell in love there. Whatever the details, Bernardo had been living and
working in the United States for eight years when Liberata arrived in 1913.
Within a year their first child, a son named Michael Angelo, was born.

The family settled in Donora, where Bernardo found work at the metal factories as a trackman.

Just five feet, four inches tall and 145 pounds, the hazel-eyed Di Sanza performed brute work. He changed railroad ties, built new track, and loaded and unloaded freight throughout the Donora plants. There were more than a hundred miles of railroad track within the plants' narrow confines. The track, owned by Donora Southern Railroad, a subsidiary of US Steel, carried materials and supplies from one building to another, molten iron from the blast furnaces to the open heart furnace for processing into steel, and molten slag from the steel and zinc plants to assorted dumping grounds along the river and the hillside north of town. Di Sanza became intimately familiar with every mile and finished his career there as a track foreman, retiring just a year before the deadly fog. By that time he had become bald, a portly spark plug of a man whose body, even at sixty-seven, remained vigorous.

The factories in Donora were filled with such men—sinewy, robust, courageous men who took on some of the most physically demanding work imaginable. Intense heat was but one of the many challenges for workers in a steel mill or, especially, a blast furnace.

Blast furnaces required workers who could deal with intense heat and who also possessed the wherewithal to avoid carbon monoxide, an ever-present gas in certain areas of the furnaces. The gas is normally odorless but when combined with other waste gases in a furnace would take on a particular odor. Charles Rumford learned to detect that odor early in his steelmaking career. Rumford described in his book, *Steel: The Diary of a Furnace Worker*, a conversation he had with an Italian co-worker after he started working at a blast furnace near Pittsburgh:

In a minute or two, we stepped out on the platform on top of the furnace. Adolph looked at me and grinned. "You smell dat gas?" he asked.

I nodded. He referred to the carbon monoxide that I knew issued from the top of all blast-furnaces.

"You stay li'l bit, pretty soon you drunk," he said.

"Let's not," I returned.

"You stay li'l bit more," he continued, his grin broadening, "pretty soon you dead."

In a way, furnace and mill jobs selected the people who would perform them, rather than the other way around. Jobs that involved, say, tapping a blast furnace required workers who could tolerate extremely high temperatures for sustained periods. Anyone placed in that kind of role who couldn't tolerate the temperatures was moved elsewhere or quit. "Somebody will get hired," explained George Leikauf, professor of environmental and occupational health at the University of Pittsburgh, "and two weeks later he'll say, 'I can't take this, I'm leaving.' If you weren't strong and healthy, you probably couldn't function. If you can't function, you leave."

Little wonder, then, that mill workers, particularly those performing the most dangerous jobs, tended to be rugged, strapping young men, adventurous men, daring men who weren't afraid to get their hands dirty, their clothes filthy, their muscles thickened, their fingers calloused, and their face darkened and toughened over time from the searing heat. They were men who hauled slag, wielded mighty wrenches to repair enormous equipment, poured molten steel out of white-hot furnaces. They were exactly the kind of men supervisors looked for when hiring workers for a new zinc plant, a plant that would make the discomfort of a steel mill pale in comparison.

———

Although William Donner had planned to construct a zinc mill in Donora and had left room on the flood plain for it, by the time it was built he was barely involved with Donora and US Steel. A year after merging his Union Steel Company with Sharon Steel, in 1902, he sold both companies to US Steel, in which he still held shares. Donner then purchased Cambria Steel in Johnstown.

Now wealthy, William and Adella, along with their three remaining children—Robert, Joseph, and Margaret—moved to Shadyside, one of Pittsburgh's more affluent neighborhoods. Living within a single mile of each other in that hallowed neighborhood were Richard B. Mellon, Andrew Carnegie, Henry J. Heinz, George Westinghouse, Henry Clay Frick, and now Donner. With so many famously wealthy neighbors Adella could proudly entertain the city's elite, which she did. Often.

William preferred to work, however, and when he wasn't working he wanted to sit quietly at home. William's homebody tendencies and need for privacy clashed with Adella's social side and need to flaunt her new-found wealth. Eventually the marriage crumbled. Adella is said to have begun an affair with a "prominent young business man" sometime around 1904, and when William found out he filed for a legal separation. Adella, who soon moved to Cleveland, accepted two hundred thousand dollars in return for granting William a divorce, along with full custody of their children, on the grounds of "neglect of duty."

Donner would remain single for five years before marrying his second and last wife, Dora White Browning. Dora was born January 26, 1884, in Camden, directly across the Delaware River from Philadelphia, and was twenty years younger than William. Dora's first husband was the son of William Rodgers, who had captained one of the steamers involved in the Homestead strike in July 1892, one of the longest, most rancorous labor clashes in US history. Dora and John Norwood Rodgers had married in 1903, two years before a fatal mishap tore the union asunder.

Young John was killed when a cable snapped on the dock he was stand-ing on. The blow sent him flying several feet onto the deck of a nearby boat. "He alighted upon his head," read a newspaper notice about the inci-dent, "and when picked up was found to be unconscious and in a very serious condition. He was hurried to the hospital in an ambulance, imme-diately after which an operation was performed upon his skull. The hospi-tal physicians have little hope for the patient's recovery." Rodgers died six days later, on January 14, 1905, at Homeopathic Hospital of Pittsburgh, now Shadyside Hospital. He was twenty-eight years old.

Dora must have been crushed. She was just twenty-nine herself and was now not only a widow but also the lone parent of two small children, Dorothy and Katherine. It would be four years before Dora married Wil-liam Donner, whose own marriage had only recently dissolved.

The morning of Dora and William's wedding, March 27, 1909, was a chilly one in Pittsburgh, with temperatures refusing to climb above freez-ing until around 11 a.m. It was around then that the bride, her family, and her friends realized that the groom was nowhere to be found. The cer-emony was scheduled for noon. Unbeknownst to Dora and the others, Donner was still at his office in downtown Pittsburgh.

Lacking a marriage license, he had sent his attorney scurrying to city hall to retrieve a clerk and bring him straightaway to the office, where-upon they would all whisk themselves to the wedding venue, the home of Dora's former father-in-law, William Rodgers. Noon came and went, however, before the groom, his lawyer, and Joseph S. Werry, assistant clerk in the marriage license office, appeared at Rodgers's home. Paperwork completed and the marriage license signed, the rather peeved bride, one might assume, strode to her uncharacteristically tardy betrothed and married him.

Two years later Donner happened to meet a man named Frank Burkett Baird while on an around-the-world cruise with Dora on the steamship SS *Carmania*. Baird was a major New York industrialist and civic leader and had been a close friend of the recently assassinated President William McKinley. Donner and Baird struck up a friendship. "Both of us being manufacturers of iron," wrote Donner, "we had many mutual friends and became quite well acquainted."

The pair encountered one another several times at the renowned Shepheard's Hotel in Cairo and around town in nearby Luxor. Shepheard's had once been occupied by Napoleon's army at the turn of the eighteenth century during its invasion of Egypt and had been a favorite of the world's monied. Perhaps Donner and Baird dined on the hotel's luxurious terrace, with its graceful, stylish reed and rattan armchairs overlooking the comings and goings of busy Ibrahim Pasha Street. Perhaps they traveled together on the tram that ran directly in front of the hotel. Regardless, the two got along exceedingly well and remained friends long after the trip ended. When Baird was considering purchasing New York State Steel Company in Buffalo in 1914, he contacted Donner.

Donner initially showed no interest in joining the venture, but Baird persisted and finally persuaded Donner to travel to Buffalo in the fall of 1915 to discuss the idea. Donner recalled that Baird said "he would meet me upon my arrival, take me to breakfast, motor me out to the plant, take care of me during the day, and put me on the night train to Philadelphia, where I was then living. He explained that the plant was then owned by a bondholders' committee and could be secured at a great bargain." Donner eventually purchased the plant for $2.5 million and renamed it Donner Steel Company.

Donner surely followed news of his namesake town's latest addition, the Zinc Works, but it seems unlikely he had much of anything to do with its construction or operation. He had moved on to bigger adventures by then and lived three hundred miles away, on the other side of the state.

From the announcement in the early 1910s that a zinc factory was to be built in Donora through the 1930s, potential employees from around the nation, particularly those who worked in zinc smelters in the Midwest, kept arriving in Donora, looking for work. One such worker, W. Robert McCarthy, came to Donora in 1934. McCarthy had grown up in northwest Arkansas and had worked in zinc smelters in Arkansas, Texas, and Oklahoma. He and his family headed east when the smelter he worked in closed for good, just one of the many casualties of what President Herbert Hoover famously labeled a "passing incident in our national lives."

Immediately after World War I, industries previously committed to war efforts turned their attention to the mass production of automobiles, appliances, clothing, and a legion of other products needed or desired by a growing population. In 1919 the United States produced fully half of all goods manufactured in the world. Unemployment stood at just 6.9 percent.

The 1920s brought massive social and political changes. The nation's wealth more than doubled, jazz came of age, the Eighteenth Amendment banned the sale of alcohol, women earned the right to vote, new home construction soared, and for the first time in the nation's history more people lived in cities than in rural areas. Americans bought cars and appliances with abandon, many purchasing the items on installment plans. The stock market saw record growth throughout the twenties. In early October 1929 economist Irving Fisher proclaimed, "Stock prices have reached what looks like a permanently high plateau."

He was stunningly wrong. Just days later, on October 24, which came to be known as Black Tuesday, the stock market imploded. Some sixteen million shares were sold. Billions of dollars were lost. Thousands of investors went instantly bankrupt. Will Rogers, the great humorist, happened to be visiting New York City two days later, on Black Thursday, and wrote, "When Wall Street took that tail spin, you had to stand in line to get a window to jump out of, and speculators were selling space for bodies in the East River."

He was exaggerating, of course. There were no suicides from the crash that day, though the nation's suicide rate did rise significantly from Black Tuesday through the end of 1931. One Milwaukee man, Wellington Lytle, who lost all but four cents, left a suicide note in his hotel room: "My body should go to science, my soul to [Secretary of the Treasury] Andrew W. Mellon, and sympathy to my creditors."

Americans all over the country saw their savings disappear. Banks closed. Businesses shuttered. Millions of everyday Americans suddenly found themselves out of work, including zinc worker McCarthy, who had no choice but to find work somewhere. Anywhere.

McCarthy and his wife, Arvella, along with their son Robert E., headed east. They ended up following the path of an oil and gas pipeline crew. By day the men laid huge pipes to the ground, hooking them together one by one. By night they slept in communal tents. Somewhere in West Virginia in 1935, McCarthy heard about the need for workers at a zinc mill in Donora, so off he and his family went. Robert quickly secured a job, his family settling in the south end of Donora, near the blast furnaces. The second and last McCarthy child, Patricia, would experience the worst of the death fog just a week after she turned twelve years old.

WOODEN SHOES
AND AN
OATMEAL LUNCH

MANUEL RODRIGUEZ WAS A HANDSOME MAN WITH DARK, KIND EYES AND a rugged, square jaw. He had dark hair and a thin mustache curling along his upper lip. Rodriguez was born in northern Spain on January 20, 1889, to Jose and Rosa Rodriguez. He immigrated to the United States in 1910, arriving at Ellis Island on November 11 and most likely passing through Customs in just a few hours, as did the vast majority of immigrants at that time. His wife, Adelayda, pregnant when Manuel left, arrived the following year with the couple's two tiny children: Frank, two, and Armanda, five months.

The young family headed to Cherryvale, Kansas. Cherryvale was part of a mineral-rich area known as the Tri-State Mining District. The district, composed of southeast Kansas, southwest Missouri, and northeast Oklahoma, had been a key source of zinc and other minerals since the late 1870s. By 1899 Kansas was the nation's leader in zinc production.

Zinc swirled in Rodriguez's blood. He was born and raised in Castrillón, a coastal area in the center of the Asturias region of northern

FIG. 7.1. Manuel Rodriguez and his wife, Adelayda. Courtesy of Richard Lewis.

Spain, an area ripe with zinc smelters. Asturias and its neighboring region, Euskadi—together commonly referred to as Basque country—had been hubs of mining and metallurgy since ancient Roman times. Rich mineral deposits had been found centuries ago in several areas of Spain, including the Basque region, where Rodriguez came from, and the Castile–La Mancha region, famed as the setting for Miguel de Cervantes's epic tale *Don Quixote*.

Rodriguez might well have gone to work at one of the local smelters when he came of age, but if not, he almost certainly knew zinc workers and their families. He and his family had followed the path of a large number of other Asturians who had already found work in Kansas zinc smelters. Between 1900 and 1924 thousands of Asturians immigrated to the United States to find work. Most of them were skilled mine workers or metallurgists seeking opportunities in the New World, predominantly in California, Florida, Indiana, Kansas, Missouri, New York, Ohio, Pennsylvania, and West Virginia. Rodriguez found work at the Edgar Zinc smelters on Cherryvale's north side.

Manuel and Adelayda had four more children in Kansas: Mary Carman, Oliva, William, and the youngest, Joseph, whom everyone called Jiggs. All six children were raised in Kansas and also married there. As the century wore on, though, the family found life increasingly difficult. After World War I the need for zinc plummeted. Prices dropped, and profits tumbled. Zinc smelters throughout Kansas closed one after another

Just as the McCarthy family had survived the Great Depression and found themselves in Donora, so too had the Rodriguez family. In 1937 two of the Rodriguez children, Frank and William, along with Mary Carman's husband and Oliva's husband, hopped into a car and drove to Donora, where they found immediate work at the zinc smelter. They probably stayed initially with friends or relatives who had already moved to Donora. Eventually all four of the men and their families found homes in the north end of town. The rest of the clan, including Manuel and Adelayda, left Cherrydale for Donora a year later. Manuel and Jiggs found work alongside Frank and the others at the zinc plant, making zinc production a true family affair.

The families found comfort and camaraderie in Donora's north end. Spaniards formed a significant portion of the labor force of the Zinc Works, the northernmost mill on the river. Many other Basque immigrants lived in the town's north end. Manuel and Adelayda lived about a block from Mary Carman, who lived just a quarter mile from the Fourteenth Street gate, which led directly to the Zinc Works, where Mary Carman's husband, Harry Lewis, worked.

The Zinc Works, built in two phases and completed in 1916, used a Belgian process for producing zinc, which involved heating zinc ore in hollow tubes called retorts. The Zinc Works contained a total of 3,648 retorts placed horizontally in huge grids in the smelter, the main section of the zinc plant. Harry Lewis was one of many furnacemen working with the retorts.

Donora used horizontal retorts to extract zinc from zinc ore that had been mined and then crushed into a powder called zinc concentrate. Each retort was a cylinder fifty-two inches long, with an interior diameter of eight inches. Closed on one end, the retorts were held in a row of metal stacks. Each stack held two retorts, side by side, in each of four rows. The retorts would be slid horizontally into the stacks, with the open end facing out.

Those stacks had been built into enormous two-sided furnaces that would heat the closed end of the retorts. The Donora smelter held nine of those furnaces—six with 912 retorts each and three with 608 retorts each.

The process of extracting pure zinc in a retort began in a different building in which zinc concentrate was mixed with powdered coke and then pressed into bricks. The bricks would be sent to the smelter, where furnacemen would pull a retort from its rack and fill it with the bricks, now called the charge. The men would seal the opening with wet coal and load the retort back into the stack. The process would be repeated over and over, the furnacemen moving along each rack and charging each retort it held.

The furnace would heat the retorts to about 3,600 degrees Fahrenheit, at which point two chemical reactions would occur. First, oxygen molecules, attached to zinc molecules in the charge, would be released. The oxygen molecules would combine with carbon molecules in the coke and form the gases carbon monoxide and carbon dioxide. At the same time zinc would turn into a vapor. A condenser at the front of the retort would collect the vapor and turn it back into a liquid, which was then collected by workers.

Lewis and other furnacemen generally wore goggles and heavy gloves to protect against zinc spray. Even a tiny amount of moisture dropped into molten zinc could unleash a fine cloud of nine-hundred-degree zinc into the air. Errant drops of sweat from a worker's forehead could lead to severe burns to his face, neck, or any unprotected area. Burns were such a common injury that only the worst burns received medical care. Even workers with severe burns were often forced to return to work the next day.

Every eight hours a furnaceman would tap each retort, breaking the coal plug with a rod and allowing whatever molten zinc had collected from the burning of the charge to pour into a kettle. The kettle would be moved along its tracks in front of a panel of retorts, collecting molten zinc at each retort. Teams of furnacemen throughout the smelter continuously collected molten zinc in kettles. When a kettle was full, it would be poured into a mold, which would be transferred elsewhere in the mill for processing.

Each charge in a retort took about twenty-four hours to release all of its zinc, at which point the retort would be removed and cleaned. Another charge would be added, the retort would be placed back into the racks, and the process would begin anew.

Donora's retorts ran twenty-four hours a day, seven days a week, all year long, holidays included. By 1920 the Zinc Works was producing an estimated 34,000 to 41,000 tons of zinc per year, more than any other zinc plant in the world. Keeping all of those retorts working required an enormous amount of zinc concentrate. At one point Donora's retorts devoured about 70,000 tons of zinc concentrate a year.

The smelter procured some of the zinc concentrate it needed from mines in the Tri-State Mining District, where Robert McCarthy had worked. The largest supply, however, came from a desert town in New South Wales, Australia, called Broken Hill, some ten thousand miles away. Broken Hill was a mining town in the Australian Outback, about 270 miles northeast of Adelaide and 600 miles west of Sydney. It was there in 1883 that Charles Rasp spied an odd outcrop of land he believed held significant mineral deposits. The land was on a sheep ranch named Mount Gipps Station, so Rasp enlisted its owners, plus a few other investors, to mine the land. Two years later the men formed what became known as the Syndicate of Seven, consisting of Rasp, Mount Gipps manager George McCulloch, sheep overseer George Urquhart, bookkeeper George A. M. Lind, ranch hand Philip Charley, and two contractors, David James and James Poole. They each invested seventy pounds, quite a sum in those days.

The Syndicate of Seven found on that outcrop a sizable deposit of zinc ore, as well as deposits of lead and silver. Within two years the syndicate had formed the Broken Hill Proprietary Company, or BHP, which later became BHP Billiton. BHP remains one of the largest mining companies in the world.

One of BHP's earliest successes, and the main reason for its rapid growth, was its invention in 1901 of a process for extracting zinc concentrate from raw zinc ore. The process, called flotation, removed fluorine-containing minerals, such as fluorspar, from the ore. Fluorine-based minerals can damage the lining of furnaces that turn zinc concentrate into highly purified zinc, so finding a process to remove it before sending the concentrate to a zinc smelter made Broken Hill's products extremely valuable.

Over the years Donora's zinc factory continued to use zinc concentrate from Broken Hill, gradually blending in concentrates from other mines as well, including the Davis-Bible mine in east Tennessee, three mines in New York, and various mines in Montana, New Mexico, and Washington State. No single supplier could provide enough zinc concentrate to satiate the intense hunger of Donora's roasters, kilns, and retorts.

Heat could hardly be avoided by workers at the retorts or operations that supported the retorts. Like other laborers working on the smelter floor, Harry Lewis wore long underwear all year long to insulate him from the heat. Gloves and long-sleeve shirts were standard wear as well. The floors near the retorts would become so hot that they could melt leather shoes, so Lewis wore wooden clogs. He used a rope as a belt; a metal buckle would have scorched his abdomen. He couldn't wear Levi's jeans, lest the rivets leave little round burns on his hips and buttocks. The skin on his face was constantly ruddy and rough from the intense heat.

Regardless of where someone worked in a smelter, heat was omnipresent. A smelter consisted of four main processes, each occurring in its own plant, and heat was central to them all. In the first process, called roasting or calcining, zinc concentrate is heated to one thousand degrees Fahrenheit in a roaster, a series of round kilns built one atop the other, each with a fine grate at the bottom. Zinc concentrate was dumped into the top kiln, where it met extreme heat. The heat caused the chemical bonds that adhere zinc and sulfur in the concentrate to erupt, separating the two and releasing them into the air inside the kiln. The sulfur quickly attached to oxygen and turned into sulfur dioxide, a gas. The zinc also bound to oxygen and turned into zinc oxide, a solid.

Now a smaller solid than the original concentrate, the zinc oxide dropped through the grate into the basin below. A set of metal-alloy arms, like agitator arms in a flour sifter, were pulled through the basin by a motor attached to a vertical central axle. As the axle turned the arms rotated and stirred the zinc oxide mixture so that it burned evenly. That mixture underwent the same chemical changes, and the resulting zinc oxide mixture dropped to the next basin, and so on.

That repeated process purified the concentrate, now called roast. The roast was then moved to a sinter plant. Sintering involved mixing the roast with a high-carbon form of coal, called anthracite. That mixture was placed on a convoy of iron pallets moving slowly along a track. Metal grates on the bottom of each pallet allowed heat and gases to pass through the mixture. Like bread on a conveyor toaster at a roadside diner, the pallets were heated as they passed along the tracks. Most of the sulfur and other impurities in the roast were now burned off into an ultrafine dust, or fume. The fume and much of the other gases generated throughout roasting were vented to another part of the plant. The material left behind consisted of refined, but still not pure, zinc oxide, which was then crushed and sent to the mix house.

Mixing, the third process, was by far the dirtiest. Workers in a mix house used pug mills, or sometimes pan mills, a similar machine, to mix the highly purified zinc with fine anthracite coal. A pug mill employed a rotating shaft covered in metal paddles to mix zinc and coal into a seamless blend. The blend was then sent to the retorts, the final step in zinc processing.

Art Larvey, a consulting chemical engineer in Pennsylvania, recalled his days working in a mix house. "If I was in the mix house for a full day on a project," said Larvey, who is White, "I'd come out looking like Othello. Smoking was absolutely forbidden in there because of the potential for a dust explosion from the fine coal. If there was a hell hole in a retort plant, the mix house was it."

If the mix house was a hellhole, the roaster was hell itself. On top of having to endure intense heat and constant swirls of black dust, roaster workers also had to withstand the piercing odor and sense-stinging qualities of two noxious gases, sulfur dioxide and sulfur trioxide. Sulfur oxides were a constant presence in the roaster. Not only were concentrations of the gases highest near the actual roasters, but the gases also tended to accumulate in stagnant areas of the plant, such as corners and offices. The gases could reach dangerously high concentrations anywhere in the plant, especially during winter, when doors and windows were often closed.

Both gases can burn the lining of the nose, throat, and lungs, causing cough, pain, and shortness of breath. Even short-term exposure can cause eye irritation, dizziness, headache, nausea, and vomiting. In high concentrations the gases can provoke a life-threatening buildup of fluid in lung tissues, called pulmonary edema. Someone with pulmonary edema struggles to breathe and, without medical care, eventually suffocates. Plant managers and laborers alike suffered the searing effects of sulfur oxides, principally sulfur dioxide, the more dominant of the two gases inside the plant.

In other Pennsylvania zinc smelters, probably in Donora as well, new hourly employees would be assigned to the roaster. The longer a worker stayed, the more seniority he built and the greater his chances for bidding out of the roaster to a plant with less destructive conditions. "Most workers bid out of the roasters as soon as they could," Larvey explained. "Many quit within a few months of being hired. Attendance was the worst in the entire smelting complex, and the accident record, due to constant employee turnover, was terrible."

Roaster workers in the early years of the Zinc Works tried to protect their lungs from damage by putting a wet bandana between their teeth and breathing through their mouth, rather than nose. Later, respirators with special filters to block the inhalation of acidic gases became available. The respirators helped prevent at least some of the damage caused by the gases. Even plant managers working inside enclosed offices were sometimes forced to wear respirators to be able to keep working.

The most difficult and dangerous job in a roaster, though, and arguably in the entire mill complex, belonged to the roaster helper. Most zinc roasters were run by a number of teams working in shifts, each team working one roaster station. A typical team consisted of an operator, an assistant operator, two boiler attendants, and two helpers. The main job of the helper involved raking the roast, similar to stirring meat in a massive frying pan. The stirring kept the mixture evenly heated. First, one of the helpers would open a door to the roaster. With temperatures inside ordinarily exceeding 1,800 degrees Fahrenheit, a blast of scorching air would strike the helper like an unexpected slap across the face. The helper would grab a long hoe-like tool and wait for one of the metal arms in the kiln to pass. Then he would shove the hoe into the inferno and rake the mix as hard and as long as he could.

When he couldn't take it anymore—whether from fatigue, heat, gases, or all three—the second helper would take over. When that worker couldn't take it anymore, the first helper jumped back into the fray. "It could take one or two rotations of the arms to clean things out," said Larvey, "or it could take a lot longer. Sometimes you had to stop the arms and cut off the feed to get back on track. The helper's job took strength, coordination, endurance, and fast reflexes." Roaster helpers worked only a few hours a day, but even those few hours must have seemed unimaginably long.

Smelter workers faced a cannonade of toxins every day. The list of gases and fumes emanating from the various parts of the smelter reads like the index in a chemistry book: sulfur dioxide and trioxide, carbon monoxide and dioxide, cadmium sulfide, cadmium oxide, lead sulfate, arsenic, fluoride, and a host of lesser elements. Those toxins either escaped into the air inside the plants or were released into the atmosphere through the mill's smokestacks, and typically both. None of those toxins were harmless, and in all probability everyone—from common laborers to managers to business owners—knew it. After all, everyone was breathing the same dusty, stinging, malodorous air.

Most workers in the zinc smelter worked just four hours a day, rather than the typical eight-, ten-, even twelve-hour days worked by those in other Donora plants. Four hours proved the longest anyone could tolerate the toxic dust and gases in the mill without suffering injurious effects. Furnacemen like Harry Lewis were directly exposed to "blue powder" coming out of the retorts.

Horizontal retorts of the kind used in Donora tended to leak zinc vapor into the air. The vapor immediately turned to a fine, bluish powder that floated everywhere inside the mill. Blue powder alone could cause sickness, but it was even more toxic when combined with cadmium oxide gas. Cadmium occurs naturally as cadmium sulfide, a yellowish coating on zinc ore. When the ore is heated, the cadmium sulfide coating reduces to

FIG. 7.2. A worker points at a panel of horizontal retorts in the Donora zinc smelter, as toxic zinc fumes pour out of the retorts. Courtesy of Donora Historical Society.

cadmium oxide gas. When inhaled the gas can cause devastating, possibly fatal lung damage.

Workers in the retort areas of the smelter, particularly those like Lewis who worked directly with the retorts, commonly experienced what experts now call metal fume fever. Also called Monday morning fever, zinc jitters, and zinc shakes, metal fume fever occurs from the inhalation of zinc oxide, cadmium oxide, magnesium oxide, or oxides of certain other metals. The exact mechanism by which those gases cause illness remains cloudy, but the effects are crystal clear. After a few hours at the retorts Lewis would have developed fever, headache, shortness of breath, or nausea. His joints would have ached. He would have had chills and felt as if he had the flu. In fact, the symptoms of metal fume fever and influenza are almost identical. The main differences lie in how long each lasts.

The nickname "Monday morning fever" arose because workers experiencing the condition during the week would feel better during the weekend, when they hadn't been breathing the fumes. Monday would come, the worker would return to the smelter, and the symptoms would return. Whereas someone might have the flu every so often, zinc workers would have the "flu" every workday of every year, year after year.

Common cures for metal fume fever at that time included drinking a mixture of water, milk, ice, oatmeal, and whiskey while standing outside in the "fresh air." Lewis learned to keep metal fume fever at bay, at least to an extent. "To keep himself hydrated and energized on the job," said Rich Lewis, "my grandfather would only take water and oatmeal in his lunch bucket for his break at work. Other foods would make him sick when dealing with the extreme heat and stress."

Besides feeling unwell most of the time, zinc workers also tended to experience long-term health effects. Zinc and cadmium oxides inhaled in large quantities over an extended period can lead to permanent lung damage. Oxide gases can also cause permanent kidney damage, and they greatly increase the risk of developing lung cancer, a common ailment among coal, zinc, and steel workers.

"Those guys were inhaling huge amounts of high doses of those toxic chemicals," said occupational health professor George Leikauf. "Cadmium was all over the place, and it's a really nasty carcinogen. And then you throw in cigarette smoke on top of that—and pretty much everybody smoked back then—and your load of cancer-causing chemicals is amazing."

8

MR. EDISON
ARRIVES

DONORA'S POPULATION IN 1916 STOOD AT A BIT MORE THAN 10,000, with the factories employing 6,200 men and 300 women. Officials expected that with the operation of the new zinc smelter, plus a new rod mill being built, the population would exceed 20,000 by 1920 and perhaps go as high as 25,000. At the time there simply weren't enough houses to go around. An article in the *Pittsburgh Post-Gazette* in August 1915 put the issue into stark relief: "The action of the United States Steel Corporation in building a new smelter plan here, transforming in three weeks a field of fifty acres from an ore dump to a hive of industry, and bringing 1,400 men as workers, has produced a housing and boarding condition in Donora that may be hard to cope with."

Officials found whatever temporary housing they could for workers and used local restaurants to serve lunches, but as the *Post-Gazette* reported, "accommodations are proving entirely insufficient. Some of the 1,400 men have found lodging and boarding in Webster, across the river from Donora, and many are traveling to and from Charleroi, Monongahela, and other points. Some are even traveling daily from Pittsburgh out to their work. Donora real estate men are trying to cope with the situation." The paper described the situation as a "house famine."

Adequate housing had actually been an issue since the earliest days of Donora. William Donner, who had examined and considered a number of sites throughout the Greater Pittsburgh area, explained his rationale for choosing the Donora site this way: "The land between the Pennsylvania Railroad & the river was an excellent manufacturing site. It provided space for a larger plant than we proposed to build. We acquired the lowlands for dumping slag, and located the ore yards, blast furnaces, steel works, blooming mill, billet mill, rod mill, finishing mill, etc., for the economical handling of raw materials, semi-finished and finished products. We had ample room for future expansion."

His site might have had ample room for expanding the factories, but the area chosen for the town lacked space to house an expanded workforce. Perhaps Donner assumed that houses would be built on the other side of the hill and at the north and south ends of town, but he might have cast that assumption aside had he fully considered the area's topography and how towns tend to grow. The two important components for growth in a town Donora's size would have been a walkable main street, where businesses can flourish, and sufficient housing near where residents tended to travel each day.

Donora's McKean Avenue remains walkable to this day, and businesses did indeed flourish throughout the first half of the twentieth century. Houses were indeed built north and south of town, and a few were built over the top of the hill. However, the density of houses in Donora, even in 2022, drops precipitously as the distance from the factories increases. Housing contractors were simply running out of suitable land on which to build homes.

Workers wanted and needed to be close to the factories. So when potential mill employees came to Donora, most looked for housing within walking distance of the plants. If they couldn't find a house or couldn't afford one yet, they faced few other options.

Some found shelter by renting space in boardinghouses, staying in hotels, or finding whatever space was available in tents, churches, ethnic clubs, or private homes. The Union Improvement Company purchased, sometime between 1917 and 1920, the Indiana Hotel at Sixth and Meldon to house workers, but it wasn't enough. "The oft-told story," explained

Brian Charlton, "of three steel workers sharing a single bed by sleeping in shifts did not originate in Donora but was routinely practiced there." Charlton calls those shared berths "hot beds," because "the beds never got cold."

In seeking solutions officials of American Steel & Wire learned that concrete houses were being built quickly and inexpensively in New Jersey. Those homes had been designed and built by the famed "Wizard of Menlo Park." Thomas Alva Edison was arguably America's greatest innovator, having by that point already invented the phonograph, an improved telegraph system, the motion picture camera, alkaline storage batteries, the first commercially successful light bulb, wax paper, electric pens, talking dolls, and mail-order subscriptions. During his lifetime Edison received nearly 1,100 patents, including several for concrete products.

Edison had been experimenting with concrete for several years before constructing a single-pour concrete dwelling in Montclair, New Jersey, in 1912. In 1899 he had started the Edison Portland Cement Company, a company that eventually failed—but not before providing concrete for Yankee Stadium in 1922. The Montclair house was different.

"The object of my invention," Edison wrote in his 1908 patent application, "is to construct a building of a cement mixture by a single molding operation—all its parts, including the sides, roofs, partitions, bath tubs, floors, etc., being formed of an integral mass of a cement mixture." Edison wanted working-class families to have access to affordable, comfortable, and stylish homes, believing that workers and their families deserved those things as much as did their wealthier employers. "I am going to live to see the day," he famously said, "when a working man's house can be built of concrete in a week. If I succeed, it will take from the city slums everybody who is worth taking."

Edison calculated that the shell for a "decent house of six rooms" could be built with "only three hundred dollars. Not just any shell, but an appealing one. We will give the workingman and his family ornamentation," he said. "They deserve it; and besides, it costs no more, after the pattern is made, to give decorative effects."

American Steel & Wire officials believed that Edison's new construction methods might quickly solve the shortage of housing for foremen and lower-level managers, everyday workers be damned. Edison had, after all, promised that his single-pour houses could be built in a week and that "it is feasible to beautify such a house far beyond anything now possible in so cheap a manner." The company decided in early March 1916 to build sixty single-family houses and twenty duplexes, enough for a hundred families, using Edison's newly patented techniques.

The homes would be built on a nearly nine-acre patch of land on the hill above the blast furnaces at the southern end of town. The area was soon dubbed Cement City, a moniker based on concrete's main ingredient, portland cement. First patented in 1824 by Joseph Aspdin, a bricklayer from Leeds, England, portland cement is a fine, granular material that, when mixed with water, binds sand and gravel or crushed stone into an amalgam that hardens over time. Aspdin named his invention Portland cement, due to the material's color, which to him resembled Portland stone, a bed of limestone on the Isle of Portland, off the coast of Dorset in southern England.

The concrete homes in Donora used a more modern form of portland cement supplied not by Edison's cement company but by Universal Portland Cement, a subsidiary of US Steel. Teams of craftsmen and subcontractors would provide all labor and project oversight. The Donora hillside was first terraced and then graded, preparatory steps for the exciting work to come.

9

WALLS
OF SLAG

"TO ALL WHOM IT MAY CONCERN: BE IT KNOWN THAT I, THOMAS ALVA Edison, a citizen of the United States, residing in Llewellyn Park, Orange, county of Essex, in the State of New Jersey, have invented certain new and useful Improvements in Processes of Constructing Concrete Buildings, of which the following is a description."

Just so begins Edison's patent application for his plan to build single-pour concrete houses. The application goes on to explain how such houses would be built. First, workers would construct a cast-iron "double wall house," which forms a kind of mold into which concrete would be poured. The form would be built in panels and clamped together "in any suitable manner, as by means of bolts, dowels, etc."

That suitable manner proved to be clips and wedges, which took less time to install than nuts and bolts. The cast-iron form proved much more cumbersome, however, and each set of panels cost twenty-five thousand dollars to build. The panels were so heavy that raising them and cobbling them together required a crane, a workforce of thirty men, and a full day of work. So rather than using cast-iron forms, construction teams in

Donora used steel panels set into place by hand. The panels were made from rolled steel and so were thinner and lighter, yet extremely durable.

Each rolled-steel form was eight feet high and nine inches wide and could be set into place and clamped together by two or three men. "Edison's panel forms made the site look like a giant Erector set," wrote historian Charlton. Within those house-size sets were placed a system of pipes to carry concrete throughout the form. The pipes were connected to a funnel-like apparatus at the top of the form. Concrete would be poured into the funnel and distributed throughout the form. When the concrete had hardened, the form was removed to be used again for the next house.

Concrete was mixed originally near the rail yards at the bottom of the hill, where it was flat. It was then transported in barrels to the job site by horse-drawn wagons or motorized vehicles. Later the concrete was mixed at the job site itself, saving time, money, and effort. From the time building commenced in August 1916 until the following August, 10,000 barrels of portland cement would be used, weighing a total of nearly 400,000 pounds. The cement would be mixed with 5,400 tons of sand and 6,500 tons of slag from Donora's steel mills as the aggregate.

Typical mixtures of concrete used crushed rock to give long-lasting strength and rigidity to a structure. Rather than carting thousands of tons of rock from quarries outside of town, however, builders used slag that had been cooled and broken into pieces. Slag had been commonly mixed with cement for building roads and railroad beds and had also found uses in sand blasting and in roofing shingles. The steel mills in town produced untold tons of slag, so acquiring the chemically inert nuggets for Cement City seemed practical and cost effective.

The final ingredients in the concoction consisted of two elements: gypsum, a mineral essential for controlling the speed at which cement hardens, and bentonite, an absorbent clay used to suspend slag uniformly throughout the concrete. Bentonite clay was instrumental to Edison's pouring process. When Edison first proposed pouring concrete houses, many construction experts recoiled and said that his process was impossible. "It was thought," wrote Matthew Josephson in his biography of Edison, "that the heavier components of the concrete mixture, made of sand, cement, crushed stone, and water, would tend to separate out or settle in the molds unevenly." The experts underestimated Edison's genius.

Edison had created, then patented a new formula relying on bentonite clay to prevent aggregates in the mix from settling to the bottom, thereby allowing the concrete to flow evenly throughout the form. He also used more gypsum than was typically used. Edison's experiments had shown that additional gypsum increased curing time, which would allow workers to more efficiently craft stairs, windowsills, and other embellishments required for the houses. Together, all of that portland cement, sand, slag, clay, and gypsum provided a staggering 324,000 cubic feet of concrete, 3,375 cubic feet of which would be used for each Cement City house.

Home construction moved slowly over the first few months, much too slowly to keep to American Steel & Wire's extremely tight schedule. As it turned out, the winter of 1916–17 was atypically cold and wet, which prolonged curing times for the concrete and interrupted the pacing of all aspects of Edison's complicated construction process. Workers also had more difficulty than expected in learning how to build concrete homes in such a unique way. Last, the general manager of the Cement City project, Louis Brandt, had underestimated the number of workers needed to keep the project on schedule.

Brandt was an experienced engineer from Pittsburgh who had managed the construction of the grandstands at Forbes Field, very nearly the nation's first all-steel-and-concrete stadium. (Shibe Park in Philadelphia, the actual first steel-and-concrete stadium, opened April 12, 1909. Much to the chagrin of Pirates fans today, Forbes opened just two and a half months later, on June 30, 1909. Forbes was this close to being first. Home of the Pittsburgh Pirates from 1909 to 1970, fans sitting on Brandt's grandstands witnessed Babe Ruth's last three home runs and Bill Mazeroski's famous 1960 World Series–winning homer.) Besides lending his skills to a historic ballpark, Brandt had also managed housing construction jobs in Donora and in nearby Clairton, so his selection as general manager for the Cement City project must have seemed ideal.

Unfortunately, with the winter's bitter cold, Brandt's failure to bring in enough workers, and the complexity of Edison's system, only fourteen houses had been completed by the spring of 1917, putting the project well behind schedule. So US Steel contracted with two other companies for additional men and equipment. The boost helped, and construction galloped along so well that the full complement of eighty homes had been completed by that November.

———————

When Cement City first opened, renters could choose from eight styles, all modeled after the Prairie School style of homes made famous by architectural visionary Frank Lloyd Wright. They were rectangular, two-story houses with long and straight lines, a simple and appealing design. A Prairie-style home, commonly called a Prairie Box or an American Foursquare, was one of the more popular styles from the mid-1890s to the late 1930s.

Prairie-style houses could be built so easily that Sears, Roebuck, and Company, the retail giant, sold complete home-building kits through its catalog from 1908 to 1940. Each kit, sent by rail, provided a seventy-five-page book of instructions and all supplies, from "first-growth, top-grade southern yellow pine" framing members to solid maple flooring for the kitchen and bathroom to every nail and screw needed to hold everything together. Each kit contained a staggering 10,000 to 30,000 individual pieces, depending on the style chosen. Over the 32 years the company offered home kits, customers were able to choose from more than 370 styles, many of them variations on the original Prairie-style design.

Cement City homes each had a basement and porch. Some even had wooden pergolas attached to them, though none remain today. Each house came originally with a lilac bush and either a red or white rosebush, courtesy of the Union Improvement Company. Houses came equipped with a gas-fired furnace, each of which was swapped for a coal furnace in 1924, because coal at the time was cheaper than gas. Many owners in the 1940s switched back to propane gas, which had become cheaper than coal.

———————

As long as the Union Improvement Company owned the homes, residents were required to follow several stringent rules. They couldn't hang anything on the walls, for instance, and they had to agree to whatever interior paint color or wallpaper pattern the company had chosen. Union Improvement Company routinely painted or wallpapered each house every two or three years and maintained the entire neighborhood,

FIG. 9.1. Pedestrians walk along Chestnut Street, with the recently completed Cement City in the background, November 1917. Courtesy of Donora Historical Society.

including mowing, shoveling, fence painting, street sweeping, and hedge trimming.

Following Donner's original vision for a paradisaic town, the company installed tennis courts, playgrounds, gardens, picnic areas, and even a baseball field to keep employees contented. Security measures maintained the exclusivity of the little town within a town. Guards monitored all visitors, keeping a particular eye out for children who might wander into the pseudo-compound to make use of a playground or ball field. One resident, Nicholas Uhriniak, recounted, "When I was a kid you just didn't walk into Cement City to play. Someone who lived there had to invite you." Even baseball legend Stan "The Man" Musial, born and raised in Donora, apparently avoided the area. One resident remembers that when Musial visited Donora years later and talk turned to the town's most exclusive neighborhood, The Man remarked, "Cement City was where the rich kids lived."

————————

Cement City, which stands sturdily today, was in all respects a marvel of engineering. Eighty houses were built from start to finish in about fourteen months, an average of roughly one house every six days. By comparison a study by the US Census Bureau indicated that construction of a typical single-family house in 2014, being built for immediate sale, took a little over six months from permit to completion.

The cost of building Donora's concrete homes, however, proved excessive. Before construction could begin, contractors had to purchase four sets of the steel molds, each set costing $12,000, an amount equal to $241,000 in 2022. The average construction cost of each house, not including the startup expenditures, totaled $3,386.44. Edison had envisioned selling each house for $1,200. Had his vision been carried through, the project would have lost nearly $175,000.

The Union Improvement Company chose not to sell the new houses but to rent them. The entire development had been designed to house mill foremen and middle managers, most of whom had come from Scotland or Ireland. Laborers and other lower-level workers commonly, though discreetly, referred to Cement City supervisors as "silver-check men" for the color of their identification buttons. Foremen and middle managers wore ID buttons with their photo over a silver checkered background. Laborers and those in nonsupervisory roles wore buttons with their photo over a solid-color background. Laborers typically looked at Cement City residents as Stan Musial did, as the rich people of Donora, though they were anything but.

Mill officials who rented half of the smallest duplexes paid $22.50 a month, while those who rented the largest single-family units paid $42.00 a month, with the other units falling somewhere in between. In comparison, a New York City apartment in the 1910s averaged about $40.00 a month.

In 1934 the Union Improvement Company transferred all Cement City deeds to American Steel & Wire, which continued to rent units to mill foremen until 1943. At that point the deeds were sold to John Wilmer Galbreath, a real estate titan and horse breeder who owned the Pittsburgh Pirates. Galbreath would go on to build a forty-one-story tower at 525 William Penn Place in Pittsburgh, the third-largest building in the city. After purchasing the Cement City deeds Galbreath immediately began selling units, usually for about $3,300 each, close to the original cost of construction.

Cement City helped ease Donora's housing crunch. Temporarily. As the years passed, however, and especially after the houses were being sold rather than rented, the drawbacks of living there began to greatly

FIG. 9.2. An identification badge for nonsupervisory workers, belonging to Manuel Rodriguez. Courtesy of Richard Lewis.

outweigh the benefits. The houses were indeed sturdy. They were, and remain, fireproof. They proved highly durable. They maintain even temperatures and, without much effort or cost, stay warm in the winter and cool in the summer. But they are not without shortcomings.

"Nothing is easy [to fix or repair] in one of these houses," explained long-time Cement City resident Bob Turnbull. "Simple things turn into major projects. Some contractors just won't work on them. Some will ask, 'Is that in Cement City?' Oh, maybe they try it once, but most say, 'never again.'"

In some of the houses with lath-and-plaster walls, wooden studs are off-center. Cement City resident Glenn Howis, a former engineer for American Steel & Wire, said, "It's hard to get concrete perfectly square, plumb, and level. Those old-timers had to custom-fit everything, like the cabinets and woodwork, without modern power tools."

Owners of typical wood-based houses can tear down a wall here or there to change a room's dimensions, add a closet, or install a shower. Owners of a concrete home, however, can do none of those things without great difficulty. Nine-inch-thick concrete walls cannot easily be torn down. Even adding a door can prove enormously costly and time-consuming. Nicholas Uhriniak drilled, busted, and ripped two openings in the concrete in his daughter's house to make doorways to turn her duplex into a single. "It took three days for each one," he remembered. "When we got to the basement [to drill a doorway], we said forget it."

When Cement City was built few residents anywhere in town owned cars; they walked everywhere. As the century progressed, though, more and more Donorans purchased vehicles. Few houses in Donora have driveways, even today, and Cement City is especially devoid of them. Houses there are so crowded together that driveways cannot be made between them. Garages are completely out of the question. So residents park on the narrow streets, making passage of emergency vehicles, snow removal, and parking for visitors challenging.

Still, the homes in Cement City stand as testaments to human ingenuity. Even though Edison's single-pour houses proved too complicated and costly to expand the concept beyond a few isolated experiments, concrete was used in constructing numerous developments nationwide, though not through single-pour processes. Concrete housing projects were developed in Bridgeport, Staten Island, Akron, Duluth, and Gary, Indiana, home of 110 units, the largest such development in the country.

Cement City has served as a live classroom for engineering students in the area. During its construction the Carnegie Institute of Technology, now part of Carnegie Mellon University, sent students taking masonry classes to Donora to learn about this newfangled process of pouring concrete. Engineers and engineering students traveled to town to view the construction and completed homes, and an article about the development was featured in a 1918 issue of *Popular Mechanics* magazine, complete with five photos of Cement City in various stages of construction.

Perhaps most important, Cement City crystallized Edison's belief that shoddy, subpar housing damaged the spirit of the common man, the laborer, the hired hand. He possessed from an early age an urge to give the poor and working class a better life. He had promised that if his concrete homes succeeded, he would take no profit from them. Perhaps Edison's social conscience had come partly from his father, who had been exiled from Canada due to his political activism. Perhaps he felt a kinship with the underdog from having lost much of his hearing from repeated ear infections as a child. His visits as an adult to the ghettos of Manhattan probably bolstered that connection.

Whatever the origin, Edison's public spirit and unstoppable creativity gave tiny Donora a neighborhood that has since become a historical landmark. He also bequeathed to the town a sense of pride that even a deadly fog could not erase.

PART III

WORKING THE MILLS

10

TRANSPORTING
TREASURES

BORN TWO YEARS SHY OF A NEW CENTURY, HOWARD HART CAME OF AGE in the mid-1920s, a time of immense prosperity for the United States. After the so-called War to End All Wars ceased in 1918, the nation's industrial power turned full-time to the mass production of automobiles, radios, refrigerators, and a host of other consumer goods. Except for a brief depression in 1920–21, unemployment through the decade remained below 6 percent. The nation's wealth doubled during the decade. Liquor was banned by the Eighteenth Amendment, but underground liquor sales and speakeasies flourished. Jazz exploded onto the scene, teens danced the Charleston, and young flappers shocked their older, more Victorian parents with their expressions of freedom and sexuality. Italian radicals Nicola Sacco and Bartolomeo Vanzetti were wrongfully convicted and executed for murder, immigration regulations tightened, and women at long last could vote.

Howard Hart was twenty-one when the Great Depression began pummeling the nation in 1929. He struggled through as best he could, working whatever jobs he could find. Hart had married Iva Dail Coughenour in

June 1922. Their first child, Harold, born on New Year's Day 1923, died at sixteen months from pneumonia. Iva was twenty-one years old and six months pregnant with her second child when Harold died.

The conflicts in her emotions must have been dreadful, sensing her unborn child kick, one of the most thrilling feelings a woman can experience, while grieving over another child's death, the nadir of despair for any loving parent. The couple persevered, however, and within two years were parents to two healthy boys, Robert and William.

Hart found steady employment at Donora Southern Railroad sometime around 1936. He believed that the railroad was hiring conductors only under age thirty-five, and he was thirty-seven, so he lied about his age to get the job. Howard, Iva, and their now eight children were living in Newell at the time, a tiny town a few miles upriver from Howard's job at Donora Southern.

Donora Southern Railroad was an intricate rail system that used broad- and narrow-gauge trains that ran on twenty-seven miles of track within the narrow, three-mile-long foundry complex. Broad-gauge trains used standard locomotives to carry coal, coke, iron ore, limestone, molten iron, molten zinc, and heavy equipment. The trains also carried molten slag to the slag dump north of town. Narrow-gauge engines were smaller, more agile engines, commonly called dinkey engines, which moved lighter loads within the complex itself.

Donora Southern's rail system connected to a section of the extensive Pennsylvania Railroad, commonly called the Pennsy. That section ran along the western shore of the Monongahela north to Pittsburgh. The Pennsy was formed in 1846 and grew steadily over the years. By 1910 the rail system stretched from Long Island to Harrisburg, west to Pittsburgh, and north to Buffalo and the shores of Lake Erie, a range that gave customers access to the entire Great Lakes region via steamer. The Pennsy system continued to grow, and by 1950 its cars delivered supplies as far west as Saint Louis and as far north as Mackinaw City, Michigan, at the junction of Lake Michigan and Lake Huron. The linkage between the Pennsy and Donora Southern rail lines allowed the Donora Zinc Works and American Steel & Wire plants to obtain whatever supplies they needed from wherever they wanted them.

Trains in Donora ran day and night, feeding the various factories with whatever they needed. Reporter William Hard described the myriad train cars and engines traveling to, from, and around steel mills in the early 1900s:

> And there are cars, cars carrying coke, cars carrying limestone, cars carrying ladles of liquid iron, cars carrying pots of hot slag, cars carrying ingots of red steel. And there are locomotives, all kinds of locomotives, all the way from the through-freight locomotive that can haul eighty cars of coke to the little "dinkey" locomotive that looks like a toy and that hauls the steel ingots from the Bessemer and Open Hearth Departments to the rail-mill, the slabbing-mill, the blooming-mill, the billet-mill, and the structural-shape mill.

Perhaps the most common of the railroad operations was the collection and removal of slag, leftovers from the breakdown of ores to their pure metals. In the earliest years of the factories slag was dumped along the Monongahela River to raise the riverbanks and help prevent flooding. When no more slag could be dumped there, the mills found barren areas away from the mills and residential areas, the main one being just north of Palmer Park, a large recreational area originally designed for mill employees and named in honor of William P. Palmer, then president of American Steel & Wire.

When the park opened on August 21, 1921, some twelve thousand to fifteen thousand employees and family members came to play. The *Donora Works News*, a newsletter for American Steel & Wire workers, reported on the park's hugely successful launch:

> The morning of this momentous event dawned with an overcast sky and rain was threatening, but that did not deter the ardorous spirit of the picnickers who worked their way to the park. After sprinkling a little, the sky cleared up and the sun came out to cheer up those who had misgivings as to the kind of day it would be. The management had provided suitable amusements for their guests and many elderly persons, both men and women, took advantage of the occasion to surprise their younger friends in their athletic achievements. The married women had set aside their household cares and entered into the spirit of the occasion with a zest that surprised even themselves, and many a husband will hear his wife relate with pride of how she won the trophy. The kiddies too enjoyed themselves. Did you see them at the penny and peanut scrambles? And ice cream cones! There were 40,000

FIG. 10.1. Visitors to Palmer Park entered through this gate. The original pillars remain today and can be seen along Route 837, near an old Donora Southern Railroad trestle north of Donora. Photo by Bruce Dreisbach © 1920, courtesy of Donora Historical Society.

of them and they were not all taken care of by the children. And lemonade! There were 1,500 gallons consumed.

A pavilion at the park hosted many concerts in the 1930s and 1940s, including Guy Lombardo's big band and famed crooner Perry Como. The celebration also offered numerous activities, including fifty-, seventy-five-, and hundred-yard dashes, sack races, three-legged races, a skipping-rope race, a greased-pig race, and a pie-eating contest. Two other races defy explanation—a fat woman's race, won by Mrs. J. O. Neville, and a fat man's race, won by Stanley Kapton. Alas, distances for those races were not recorded.

Palmer Park was then a next-door neighbor to American Steel & Wire's slag dump. The two areas were separated by just a few hundred yards of rock and soil, but easterly breezes carrying smoke and fumes from the slag dump wafted over the park. Anyone picnicking or playing in the park breathed air laden with toxins released from the molten slag dumped next door.

———————————

Slag trains from Donora ran north on tracks along the Monongahela River and then turned west into the dump. Slag trains consisted of an engine, usually a switcher engine, and slag cars, each carrying one large ladle or a pair of smaller ladles filled with slag. The ladles, which look like immense upside-down bells, are hinged to a base, to allow the ladles to be tipped one way or the other to dump their load on one side of the tracks or the other. After arriving at a dump the slag cars would be moved to tracks laid directly along the top edge of the slag heap. Workers would release the ladles one after another and allow them to tip to the side.

The magma would pour slowly at first and then, typically, burst from the ladle like water through a crack in an earthen dam. Slabs of slag would rupture out of the ladle and explode on the ground in a spray of gray smoke and fiery sparks. Often some of the slag would hang back, clinging to the wall of the ladle, like jellied cranberry sauce clinging to the walls of a can. When stubborn slag needed help to fall a crane on the opposite side of the ladle would slam the ladle with a weighted mass of iron. The vibrations from the strike would loosen the slag enough to drop a huge, often bell-shaped monolith of molten slag. The bell would crash apart on the hill in a brilliant show of fire.

Slag trains ran every day and every night, all year long. Anyone living or traveling near the dump would have been treated to an orange glow in the sky whenever slag trains unloaded their wares. Webster residents, particularly those living on the northern hillside, would have had a nearly perfect vantage point from across the river. Clear nights and calm weather must have given those residents some spectacular shows. Webster resident Shirley Rozik recalled, "We used to watch them dump slag over that hill when we were kids. It would light up the whole hillside."

The brightest slag shows in all of western Pennsylvania, though, radiated from Brown's Dump in West Mifflin, an enormous mountain of slag just south of Pittsburgh. The 403-acre site, owned by Carnegie-Illinois Steel, a subsidiary of US Steel, began in 1913 as a small mound of earth and grew into a completely man-made landform taller than Cinderella's castle

FIG. 10.2. A string of slag pots on train cars along Brown's Dump in West Mifflin, Pennsylvania. Note that a pot of slag has just been emptied. Photo by Charles Richardson, courtesy of the University of Pittsburgh Library System Digital Collection.

at Walt Disney World. Steel mills in Braddock, Duquesne, and Homestead, as well as the Carrie blast furnaces in Swissvale, dumped about eight thousand tons of slag every day at Brown's. If the Brooklyn Bridge were made purely of slag, Brown's dump could have supplied all of it in just two days.

Families, retirees, and teenagers on dates would park their cars along the roadways and gaze in awe at Brown's spectacle. Incredulous visitors often likened the sight to a molten Niagara Falls, and they weren't far from wrong. "It looked like lava flowing down a mountain," said one former resident of the area. "Boy, those were the days. It was awesome."

Said another, "Every Wednesday our family got into the car and drove out Route 51 to the pizza shop, then drove to our spot to watch them dump the slag. My dad even bought a convertible so we could all see." Still another recalled that "in the winter, when it snowed, the hot slag would explode and hiss and smell. What a show!"

11

O! LITTLE TOWN
OF WEBSTER

LIGHT SHOWS FROM MOLTEN SLAG MIGHT HAVE DELIGHTED WEBSTER residents at night, but by day, in the nascent years of the twentieth century, Websterites had much to be delighted by. Perched along a sliver of land across the river from Donora, Webster once boasted a population of about two thousand. From the late 1800s until 1915 anyone looking eastward from West Columbia would have seen a charming town at the base of a hill, a swath of steep farmland rising over well-kept homes. "The sun rose due east above every back door and set facing the front porches," wrote journalist Scott Beveridge, a longtime resident of Webster. "Yards were separated by white picket fences, and most contained lush orchards and flower and vegetable gardens."

Founded sixty-eight years before Donora, Webster was named in honor of Daniel Webster, the celebrated New England lawyer at the height of his illustrious career. He had been a US senator for six years and had built a nationally recognized name for himself when he argued against waging war against Britain in 1812. He used his vibrant oratorical prowess over and again to foster various compromises between slave-holding and non-slave-holding states. He famously worked in concert with two other senators, John Calhoun of South Carolina and the mighty Henry Clay of Kentucky, to broker compromises throughout the 1830s and 1840s. The men fought to maintain peace between the two factions, though ultimately no peace could be brokered.

So respected was Daniel Webster in his day that a staggering number of towns, schools, streets, and even a submarine carry his name. His namesake town along the Monongahela was laid out by Benjamin Fell Beazell, a leading member of the prominent Beazell clan. The original Beazells, Matthew Basel and Katherene Smith, had each immigrated to the United States around 1760 from Basel, an ancient city along the Rhine River in Switzerland. Matthew and Katherene, unknown to one another when they embarked on their transatlantic crossing, met on the ship, fell in love, and married in Virginia shortly after their arrival on US soil. The pair eventually settled in an area now called Fellsburg, on the other side of the hill from Webster, where they built a sawmill.

The seventh of their twelve children, Benjamin Beazell, married Rebecca Fell, a native of Bucks County, north of Philadelphia, in 1782, and together they too had twelve children, the sixth of whom was Benjamin Fell. Benjamin Fell became a shipbuilder and used lumber from his father's sawmill in Fellsburg to craft his vessels. Benjamin Fell Beazell guided the town's incorporation in 1833 and, together with someone named Ford, created a map of how Webster's streets should be laid out. Beazell also donated land for the construction of the First Methodist Church in Webster. The Beazell and Fell families remained stalwart inhabitants of the area well into the twenty-first century.

In the early 1900s only a few farmhouses dotted the hillside above town. With few trees to hinder view, occupants of the homes would have been able view the entire river bend and, starting in 1901, the growing town on the other side. They would have seen and heard the sawing, hammering, and general din of men building mills. They would have watched brick buildings spring up along McKean Avenue and the Donora hillside fill in with houses. They would have seen mill trains running hither and yon, cargo trains rolling into town with coal-filled cars and rolling out empty.

And they would have known that Webster's coal mines would soon feed those new mills.

Coal mines in and around Webster were already buzzing with activity, feeding the ravenous steel mills in Pittsburgh, the Steel City. One of the oldest streets in town, Webster Hollow Road, traversed past four coal mines, and the street wasn't even two miles long. The area was so rich in surface coal, in fact, that several residents along Webster Hollow Road operated their own personal coal mines in their backyards to keep their houses warm in winter.

Pittsburgh Coal Company was the most prominent operator in town. It ran a tipple directly across the river from the steel mill. Coal tipples were wooden structures that used conveyor belts to load coal from wagons into railroad cars. Webster coal was initially transported by horse and wagon to a boatyard on the Monongahela or to a tipple next to the Pittsburgh & Lake Erie railroad track, which ran along the Webster side of the Monongahela. From there the coal would head to Steel City and soon, mine owners hoped, across the river to Donora.

Webster was large enough and important enough then to have its own bank, train station, and post office building. The town also had a large flour mill at the corner of Third Street and Webster Hollow Road. The Webster Roller Flour Mill produced King Midas flour and "feed, hay, and grain of all kinds," with "fancy Minnesota flour a specialty."

The Webster Garage sold Sinclair gasoline. A. G. Leonard ran a general merchandise store across from the railroad station. Armstrong Henry tacked shoes on horses at his blacksmith shop, and Dr. W. C. Byers, a "very agreeable young man," moved his medical practice to the town. Apparently Byers was "an eye specialist of no mean ability."

One of the more prominent men in the valley during the 1800s hailed from Webster and was buried across the river in Donora. John Gilmore, born in 1811, was a wagon-maker before the Mexican-American War and thereafter operated a steamboat to carry coal up and down the Monongahela. Gilmore married Susanna Spargo, who bore him eight children.

The family moved at some point from Webster to West Columbia, the area that would become Donora. So prominent was Gilmore that West Columbia's first school, a log cabin, was named after him. Gilmore School was built shortly after the Civil War and stood at today's corner of Fifteenth and Meldon. Besides his steamboats the wealthy Gilmore owned

FIG. 11.1. Webster as viewed from the Donora hillside, with beehive ovens (*top, center*). Courtesy of Donora Historical Society.

an eponymous cemetery on Meldon Avenue, where he and several members of the Castner family are buried, as are about forty veterans of the Civil War, including a cousin of Ulysses S. Grant. Gilmore died in 1884 and left to his heirs roughly five hundred acres of coal-heavy land.

Gilmore's hometown of Webster was one of the largest towns in the valley in the early 1900s and would have been called a "destination town" had the term existed then. During warm weather, Pittsburghers would head to the village to soak up sun along Webster's shoreline. They might visit for a day or perhaps stay overnight at one of the town's four hotels, including one at the corner of Second Street and Webster Hollow Road. That classy hotel touted a lavish restaurant with spindle-back chairs, smooth white tablecloths, and a doily-covered mantle over a central fireplace.

The town attracted men of some means, and a few built mansions there. Jacob Tomer and his wife, Harriet, built a spacious Victorian manor at the edge of a hill above North Webster. Its Queen Anne tower gave the Tomers and their ten children a glorious view of the valley below. Tomer, born in 1832, was a pharmacist before the Civil War broke out. He enlisted as a volunteer in the Third Pennsylvania Cavalry as a hospital steward, filling prescriptions for his regiment until the end of the war in 1865, when he returned to Webster to operate a large farm. He was elected justice of the peace for Rostraver in 1888 and in his later years was a member of several fraternal organizations prominent at the time, including the Independent Order of Odd Fellows, Ancient Order of United Workmen, and Grand Army of the Republic.

John Vogel, who ran the Union Hotel in town, became a citizen of some renown after serving in the Civil War. Born in 1842 Vogel began his career as a teamster in Pittsburgh. He enlisted in the volunteer army in 1861 and served four difficult years, fighting in several significant battles, including those at Yorktown, Second Bull Run, Fredericksburg, Chancellorsville, Antietam, and the famed Battle of Seven Days, when newly crowned commander of the Confederate Army Robert E. Lee rebuffed George B. McClellan's charge against Richmond in a series of battles that left more than thirty-four thousand soldiers killed or wounded.

On his return from the war the celebrated veteran built the Union Hotel and eventually owned "about a dozen of the best buildings in Webster." Vogel ran the hotel until his retirement, when he sold it to his son, John. Young Vogel immediately set about renovating the hostelry. He added a wraparound porch to the second floor, supporting the veranda with wooden pillars that gave the hotel a "metropolitan appearance." Vogel also cut two prominent bay windows in the wall of the hotel bar to add light and promote a feeling of openness in the room. The *Daily Independent* proclaimed the renovations a "credit to both the town and its enterprising proprietor."

An Englishman named Eneas Coulson owned another of the hotels in Webster. Coulson had emigrated from Northumberland in 1879, with his wife, Mary, in tow. The couple had seven children and settled in Donora sometime between 1900 and 1910. One child, Robert Jackson, would become Donora's first professional baseball player, playing from 1908 to 1914 on four different teams, including the Cincinnati Reds and the Pittsburgh Rebels.

Webster's opulent hotels, pleasant shops, and imposing homes gave the burg a charm all its own. It was a quaint, welcoming, desirable little town. For a time.

When William Henry Donner began creating steel mills across the river in 1901, Websterites paid rapt attention, and some of them were not happy with what they saw. Said journalist Beveridge, "Webster's riverboat captains, having witnessed how these giant mills had damaged other farms along their travels, immediately put their estates on the market. Anyone else with enough money soon followed their lead."

Most everyone else hung on, rooted as they were in the coal industry. From 1901 to 1915 Webster residents wanted to visit and, for some, to work in the growing town across the river. However, they could reach Donora only by rowing across the Mon, riding the Gilmore Ferry across, or, in deep winter, walking across when the river froze. Officials in both towns, as well as officials in their respective counties—Westmoreland for Webster and Washington for Donora—believed that a bridge should be built to connect them.

Proponents of the plan circulated a petition in 1902 that asked Donora and Webster residents whether they wanted a bridge to link the towns. The response was overwhelmingly positive. No longer would Webster folks need to travel to Pittsburgh, Charleroi, or Monongahela for clothing, household goods, or building supplies. A bridge would save them so much time, effort, and expense. *Yes*, they clamored, *build a bridge!*

The Washington county court, however, denied the petition based on a lack of permission from the federal government for either county to erect bridges. So Pennsylvania state representative Ernest Acheson introduced a bill in Congress that would allow the counties to proceed with the plan. The bill passed, and the counties moved forward on a new petition, which again proved highly favorable. Again the petition failed. Judge Lucien W. Doty rejected the proposal on June 25, 1904, reportedly because "it was not a propitious time to build so expensive a structure."

Once more the residents were surveyed. This time, more than a year after the Doty decision, the petition was granted, and the counties were given approval to contract with an engineering firm to build the span. Local politics flared, however, with officials in both counties quibbling with one another about contractual language and the granting of bonds to pay for the bridge.

Finally, two years later, on January 23, 1908, both counties approved a contract with the Toledo-Massillon Bridge Company, which then built the bridge in record time—just seven months from groundbreaking to grand opening. In an ironic twist, William Donner's Cambria Steel Company, located along the Little Conemaugh River north of flood-ravaged Johnstown, supplied steel for the bridge.

The Donora-Webster Free Bridge was a fixed-truss, 1,540-foot-long structure that ran from the center of Webster over the Monongahela River to Tenth Street in Donora. The bridge provided ready access to Donora's mill complex not just from Webster but from neighboring towns as well, including Fellsburg, Sunnyside, Arnold City, Pricedale, LaGrange, Indian Hill, Van Meter, and the largest town, Belle Vernon.

Fifteen thousand people attended a grand opening for the bridge. Celebrations began at 10:00 a.m. on December 5, 1908, with the first baby born in the new town of Donora, eight-year-old Rebecca Donora Castner of the influential Castner clan, cutting the ribbon. The dedication ceremony didn't last long, owing to "rough weather" on the bridge.

Webster put on a "double-ox roast," selling four thousand ox sandwiches before turning customers away for lack of meat. The *Donora American* reported that "those who were unable to get a sandwich begged a bone and many went away gnawing at a hunk or rib."

At one o'clock, two thousand people marched in a huge parade honoring industries in the area. The parade began at First and Meldon in Donora, traveled along McKean to Tenth Street, looped back through town on Thompson, crossed the bridge at Tenth Street, and finally passed before a reviewing stand on First Street in Webster. Perhaps the most romantic moment of the day came at noon, when the pastor of the Methodist Episcopal Church in Webster, R. H. Morris, joined in holy matrimony Harriet Sells Binley, of the prominent Webster Binleys, and John Charles Witherspoon, chief clerk of the Webster office of the Pittsburgh Coal Company. The wedding took place at the midpoint of the bridge, with crowds packing the bridge on either side.

And what would any bridge's grand opening be without someone leaping from it? Just after the parade passed an unidentified man climbed the railing and tossed himself over the side, sploshing into the murky waters sixty feet below. He emerged and swam to a nearby pier, clinging for dear life until he was rescued. "He seemed pleased with his experience," reported the *Daily Republican*, "but would not tell his name or talk about the incident, and hastily disappeared."

The most visually stunning moment of the day came at six o'clock, when the bridge was illuminated with "hundreds of electric lights." Festivities officially wrapped up after a celebratory smoker, held at a banquet room on the third floor of the First National Bank of Donora, at the corner of Sixth and McKean. The honored men at the smoker—men *only*—undoubtedly gathered "to smoke, drink, tell ribald stories, and probably view French postcards," said historian Brian Charlton. "A smoker was a very exclusive gathering." Each of the Very Important People invited to the smoker were handed on entry a suede pouch, the flap of which was engraved in gold with the words:

SMOKER.
Celebration Donora–Webster Free Bridge.
Constructed by Washington and Westmoreland Counties.
Pennsylvania.
December 5, 1908.

The pouch contained a swanky El Solano torpedo cigar from Cuba, which cost about two dollars then and would cost more than fifty-five dollars today. The pouch also contained two extremely popular El Symphonie torpedo cigars from E. A. Kline, a US-made El Verso Jr. corona cigar wrapped in gold foil and sleeved in cellophane, a classic half-bent pipe, a pack of "specially prepared" pipe tobacco from R. A. Patterson Tobacco in Virginia, and two Rameses II cigarettes wrapped in sealed, white paper sleeves.

Among the prestigious attendees were surely Charles W. Lutz, chairman of the celebration committee and superintendent of the steel mills in Donora; John Ailes, chairman of the reception committee and president of First National Bank of Donora; Bert Ammon, descendant of Nathan Ammon, whose land abutted Widow Heslep's property and on which the zinc smelter would be built; and Webster's own Civil War patriot, John Vogel.

The new bridge would finally connect two important towns, one whose history would be made from steel and another whose history had already been forged in coal. Hope and feelings of a new beginning must surely have radiated from Webster and Donora residents alike as they stood together on that gleaming new bridge. Hands across the river indeed.

12

ZINC IN
THE WIND

THE CONVIVIAL FEELINGS WOULDN'T LAST. WITH THE ARRIVAL OF THE zinc smelter everything on the Webster side of the bridge changed. Prevailing winds blew whatever poured forth from the zinc smelter chimneys eastward across the river toward Webster, and what poured forth were thick clouds of dust, soot, toxic gases, and particulate matter, tiny bits of solids and liquids suspended in air. Particulate matter and gases emitted from the chimneys included cadmium, sulfur dioxide, zinc, and lead, all highly toxic if inhaled.

At the height of the smelter's production nearly 30,000 pounds of zinc, 400 pounds of lead, and 332 pounds of cadmium were released into the air each day. So toxic to plants were those compounds that from the time the mill opened in 1915 until it closed in 1957, grasses, shrubs, and bushes became virtually nonexistent anywhere in Webster, and the orchards and crops that once dotted the farmland above town vanished. Nothing would grow.

Children played in weeds, dirt, dust, soot, mud, and rocks. Vegetables refused to emerge, and flowers almost never took root. Dust and soot

FIG. 12.1. Children playing in a field. Courtesy of the University of Pittsburgh Bruce Dreisbach–American Steel & Wire Company Photographs, c. 1915–17.

from the zinc smelter settled on virtually everything—curtains, window-sills, porches, sinks and countertops, chairs, couches, pictures on the wall, lamps, bed linens, nightstands, and clothing. "Every house in Webster was brown or black," said lifelong resident Bonnie Rozik. "My father would paint our house, and pretty soon the paint would come off."

Rozik's family didn't have city water, nor did most Webster residents. To clean the front porch of their home Bonnie and her sister Shirley carried water in buckets from a well behind the house. "You know how many buckets of water it took to scrub that porch?" asked Bonnie. "Fourteen."

The girls helped their mother, Iva Hart, clean the house, a nearly every-day task. "Every week the windows were washed," said Bonnie. "Thursday the upstairs was cleaned, Friday would be downstairs, and Saturday we cleaned all over again."

Just walking across the bridge could turn clothing grimy. Scott Beveridge recalled that "the slip under my mother's dress would go from white to charcoal gray by the time she walked across the Donora-Webster bridge to her job as an office clerk" at the mills.

FIG. 12.2. A young child sits on a stream bank near the railroad tracks in Donora, a town as barren of plant life as Webster. Courtesy of the University of Pittsburgh Bruce Dreisbach–American Steel & Wire Company Photographs, c. 1915–17.

Curtains were another affair altogether. Many homeowners decorated their windows with elegant lace curtains. The lace looked lovely but collected soot out of the air like a vacuum, blackening them almost overnight. Curtain-cleaning time came once a week for the Harts. Iva would take the curtains down, wash them, and dip them in a starch solution. "After that," Bonnie explained, "you had to put them on a curtain stretcher." A curtain stretcher was an expandable wooden frame with thin, slightly curved nails, called tenterhooks, sticking out from one side. Damp, wrung-out curtains would be pulled taut and stuck on the tenterhooks. Then the frame would be expanded to stretch the curtain tight. "You had to put them on the stretcher," added Shirley, "to be able to get them the length we needed, because they would shrink up." Then the girls would help their mother iron the curtains and hang them on the windows again. The entire process would be repeated the following week, not only in the Hart home but in homes throughout the valley.

Donorans didn't escape those kinds of conditions. If there wasn't an easterly breeze in the valley, all of the dust and soot from the mills settled over
Donora. Its hillside also became denuded after the zinc factory opened
operations.

Typical of many in the valley, Gladys Schempp routinely cleaned her
whole house once a week or so. Schempp, whose husband, Bill, would
play a heroic role in the smog, kept a daily diary for years. In the fall of
1948 she "cleaned [the] whole house" on September 4, 10, 19, and 30 and
October 1, 8, 15, and 25. Even on one of the deadliest days of the smog
she cleaned. Sure, the air was thick with black soot, but those curtains
weren't going to clean themselves.

Perhaps the most dramatic sign of damage from Zinc Works effluents
could be seen in Gilmore Cemetery. The cemetery, which predates by
a century the establishment of Donora itself, is located on the hillside
above Meldon Avenue, at the northern end of town. Before 1915 the cemetery functioned normally, carrying out occasional burials in family plots.
After 1915, though, the cemetery began to crumble.

Grass and shrubs died. Rain began washing away the soil covering
graves. With so much mud, and so many Civil War graves being damaged,
fewer and fewer bodies could be buried in the cemetery. The group that
owned and operated the cemetery finally decided that they could no longer bury anyone. They petitioned the tax authorities on December 14,
1924, to exempt them from the capital stock tax. They blamed the unten-
able conditions of the hillside solely on the Zinc Works. "This Plant is
directly across the railroad tracks from the cemetery," read the petition.
"The fumes have destroyed all vegetation. Deep gullies have been caused
by rains. Burials are only made in this place when the lot owners cannot
afford to bury elsewhere."

At least in Donora, people could walk up, down, and across the hillside
on streets and stairs. People in Webster weren't as fortunate. The few people who stayed in town after the Zinc Works opened lived mostly at the
base of the hill. As the grasses, shrubs, crops, and nearly everything else
growing in Webster started dying, the whole look of the town changed. It
began to appear haggard, worn down, tired. With nothing to hold soil on
Webster's hillside, rainwater eroded the slope. The soil that ran in rivers

down the hill carried with it whatever pollutants had landed on it. Residents probably did not realize how dangerous those toxins were, the toxins they breathed, cleaned, tasted, and walked in, but they would find out soon enough.

Stuart Boyd found a silver lining in all of that erosion. Boyd grew up over the hill from Webster and recalled playing on the hill as a boy. "The erosion washed away the topsoil," he said, "but anywhere there was a stone or anything hard, it left a little column, like an inch high. That sounds strange, but there were whole fields of these. We used to have a great time up there, because you could walk around and find arrowheads and spearheads, some of the most beautiful workmanship from the early natives that hunted there. Must have been tremendous hunting, because there were just so many of them."

Children might have played worry-free on the Webster hill, but adults faced a stark reality: their once lovely, charming little town and the picturesque slope above it were being decimated. Effluent from the zinc smelter, as well as from the furnaces of the steel mill, made its presence felt throughout the eastern side of the valley, from tiny Sunnyside in the north to the southernmost tip of Webster and beyond. The pollutants didn't just affect paint and plants, they also affected animals in the area, particularly farm animals that grazed on hillside pastures. Families who had depended on milk and meat from their cattle found other means of subsistence or moved to literally greener pastures.

Stores closed. Banks left town. Coal mines shuttered. Boat docks decayed. Hotels survived for a time, especially those with a bar, but eventually they too crumbled.

After the first exodus of affluent residents in the late 1910s, and then the slower, more painful exodus after the zinc smelter opened, Webster over the next few decades morphed from a destination town into an outcast hamlet. Remaining residents persevered through difficult times, hardening with each passing winter. There were few other options. Beveridge testified to the toughness of the town's residents, calling Webster "a tough-as-nails village in Southwestern Pennsylvania steel country, where differences were often settled with fists and blood after its men downed shots of whiskey chased with union-brewed beer."

To look out over the entire valley, from the 1920s through the 1960s, was to view bleak, brown slopes on either side of the river, the slopes dot-

ted with once-white, now-gray houses and huge brick buildings west of
the river belching thick, black smoke. The west side of the river would
have looked more civilized, more lived in. The east side of the river,
though, where little Webster was located, would have looked less civilized
and more like despair itself.

13

IT TAKES
A KILLING

ANGELO GIURA MIGHT HAVE HAD SOME SENSE OF THE DANGERS INHER-
ent in his job, but with his youth, the twenty-year-old probably felt invin-
cible. His supervisors, though, certainly knew the dangers. Yet neither
Angelo nor his supervisors paid any heed to those dangers, and, like Hamlet, the young man would pay the ultimate price.

Born in Italy in 1887, Giura immigrated with his parents, Michaelo
and Maria, and settled in southeast Pittsburgh, not far from Homestead
Steel Works. When he came of age the young man found work at a wire
mill in either Monessen or nearby Allenport. Short and slim Giura was
assigned to a specialized furnace, called a butt-weld furnace, as a "trough
boy," working alongside a welder. The welder would use iron pincers to
grab a flat bar of extremely hot steel as it emerged from the furnace. He
then would curl the bar's outer edges together, bit by bit, so that a rolling
machine could weld the edges along the entire length of the bar.

From the other side of the roller would emerge a round, red-hot pipe,
which would drop into an empty trough below. The trough boy, standing
at the far end of the trough, would shift a lever each time a pipe dropped

into the trough. The lever would flip the pipe onto a constantly moving conveyor bed, thus emptying the trough for the next pipe in line. The faster the welder worked to curl a bar of steel, the faster the pipes would drop into the trough and the faster the trough boy had to flip the pipe onto the conveyor. Trough boys needed to have quick reflexes, a steady hand, and an ability to pay close attention for prolonged periods.

Angelo had been working as a trough boy for about a year when misfortune struck. His supervisor had considered him "the best boy he had ever had on the job." So adept was Angelo that he took to sitting on a bench when working the lever, a move his day foreman frowned upon but which his night foreman apparently allowed.

At ten minutes before eight on the evening of January 29, 1907, a pipe dropped out of the roller before Giura had ejected the last pipe. Perhaps the welder worked a touch too quickly. Perhaps Angelo was a touch too fatigued. Ten-, twelve-, even fourteen-hour workdays were common at the time, and most workers had but one day off each week: Sunday. Whatever the cause, the new pipe struck the pipe still in the trough, shot into the air, and burned straight through Giura's body, killing him instantly. If he had been standing, it was said, rather than sitting, the ricocheting pipe would have missed him. At the least he might have jumped out of its way.

The mill soon installed iron guards along the troughs to prevent a recurrence. A labor activist named Crystal Eastman later visited the mill and the site of Giura's demise. "I have seen this guard," she said. "It seems simple and necessary. Yet it required the killing of Angelo to suggest the need to the mind of his employer."

Death and all manner of injuries were commonplace in steel mills in the first half of the twentieth century. A Bureau of Mines report in 1917 detailed the extent of accidents in Pennsylvania blast furnaces in a single year, from July 1, 1906, to June 30, 1907. The report failed to list which blast furnaces were studied, but the two blast furnaces in Donora almost certainly were on the list. The report counted 1,339 injuries and 33 deaths that year, an average of about 4 injuries per day, 114 per month. Injuries from hand labor and the use of hand tools accounted for the greatest number of injuries: 429 that year alone. Next came injuries from being struck by objects, falling, being burned by hot metal, or being injured by cranes, hoists, or riggings—a total of 511 injuries, including these:

> Two laborers were unloading heavy scrap and one allowed his end to slip. The piece in falling crushed the other man's toes.
>
> Rigger was taking strut from bridge with crane. Knocked man's leg against rail girder, amputating leg.
>
> Trestle man was holding steel wedge being driven in frozen ore, when steel sliver from wedge struck his eye. Loss of sight.
>
> Assistant foreman was on car to dislodge lump of ore frozen to side of car; lump fell and struck man, fracturing thigh.
>
> Laborer was run over and disemboweled by [railroad] transfer car.

Nearly twice as many men that year were killed while transporting materials as were killed in the actual steelmaking process. A great many of those injuries occurred to immigrants who couldn't speak English.

The Bureau of Mines study described several trends in the kind of men hired at blast furnaces over the previous years. American-born workers, for instance, found many of the most difficult and dangerous tasks "unattractive," and so refused to do them. After World War I began, the number of immigrants from England and Germany decreased significantly. Those workers had tended to be the most familiar with blast furnace operations, many of them having worked in European steel mills before immigrating. Most of them also spoke English.

With fewer experienced, English-speaking blast furnace operators available, supervisors needed to hire inexperienced workers, many of whom spoke little or no English. With so many workers now speaking so many different languages, a kind of racism reared its callous head. The Bureau of Mines report read:

> As the number of Poles, Slavs, Italians, Magyars, and men from the Balkan States entering the labor gang and working up into the furnace and other crews increased, both the older hands and recent immigrants from northern Europe declined to seek employment among them. This racial antagonism worked in other ways; strikes by employees sometimes caused a complete change in the character of the force, English-Speaking stockhouse men, riggers, and crane runners being replaced by men from southern Europe.

FIG. 13.1. A typical bulletin board at the Donora steel mill, circa 1916. The signs read, "Congratulations! We are having fewer accidents than ever before . . . And the credit is yours, men! You are making an excellent safety record! Keep up the good work!" "Don't get sore! Get first aid and resolve to be more careful." "'I forgot' has never shown profit for either worker or company." "Stop accidents at work, at home, on the street, everywhere." Courtesy of the University of Pittsburgh, Bruce Dreisbach–American Steel & Wire Company Photographs.

Only 5–10 percent of Southern European immigrants had ever worked in the steel industry, as opposed to 40 percent of British and German immigrants. Most of the immigrants at the time, roughly 70 percent, were farmers accustomed to working outdoors, often on their own. Blast furnace work, however, required men to work efficiently in teams.

"Frequently fifty percent can not speak English," the report stated, "though they commonly understand a few terms used in directing the work." Workers who could speak little or no English could easily misunderstand directions. A supervisor might recognize that an accident was about to occur and then yell a command in English, such as, "Move out of the way!" The worker who couldn't understand those words might hesitate and then be struck and injured. Combining the obstacle of not speaking English with having no experience in the "moving of large quantities of material, the handling of great volumes of poisonous and explosive gases, the use of blast at high temperatures and pressures, the pouring of fluid iron and cinder, and the operation of novel machinery," it seems remarkable there weren't more injuries.

––––––––––

As detailed as the Bureau of Mines report was, however, it wisely—from a political point of view—steered clear of suggesting that mill owners bore anything more than the slightest touch of responsibility for worker safety. "A smaller proportion of blast-furnace accidents occur through the fault of the employer," the report said, "being due to lack of safeguards, deficient equipment, and failure to give proper instructions, but even in these accidents many injuries may be avoided by pointing out the hazards and insisting on carefulness."

Even with zero safeguards, unsafe equipment, and insufficient instruction, accidents were mostly still the workers' fault. If only they had heeded the signs.

The most commonly used worker safety measure seems to have been the plethora of signs posted at entry gates and scattered around the mill site. Signs urged employees to be careful, think first, stay healthy, look before crossing, watch out for each other. One safety sign at the steel mill in Donora featured a drawing of a man with an injured finger and the words, "Don't Get Sore! Get First Aid and Resolve to be more careful." Another featured a drawing of two birds sitting side by side on a branch. One bird was labeled "Accidents," and the other, "Carelessness." In large printing were the words, "Birds of a Feather Flock Together."

No matter how many signs the mills displayed, many workers gave the signs nary a look. "It is useless to expect a Slovenian who has worked all day in the heat and glare and stress of a blast-furnace," wrote investigative writer William Hard in 1907, "to pay much attention to a danger sign, especially if he doesn't know how to read, which he usually doesn't."

Regardless, those kinds of signs clearly put the responsibility for a safe work environment on the workers themselves. Signs urging mill owners to make their mills safer for workers seem not to have existed at all.

––––––––––

A gargantuan set of studies published in 1910, together called the Pittsburgh Survey, finally called mill owners to task, muffled though the call might have been. The Pittsburgh Survey was the most extensive social survey ever conducted in the United States. It combined scholarly research with a political activism common during the Progressive Era, when social reforms took center stage throughout the nation.

Headed by editor Paul Kellogg, only twenty-eight at the time, a team of investigators compiled and analyzed statistics on a variety of aspects of life in and around Pittsburgh, from 1907 to 1908. The reports were sponsored by the Russell Sage Foundation, a nonprofit organization in the city founded in 1907 for the "improvement of social and living conditions in the United States." The survey was published in installments, some of which appeared in the Sunday edition of the *Pittsburgh Post*, and subsequently published by the foundation in six volumes.

The survey examined, in order of installments, women in the trades, work accidents, the life of steel workers, an overview of Pittsburgh and its growth, wage earning in the city, and a study of family life in a mill town, using Homestead as its example. The *Women and the Trades* volume was a particularly important document. Written by Elizabeth Beardsley Butler, a journalist and one of America's first social investigators, the report detailed for the first time wage-earning women in a major US city. For instance, about women working in the metal industries, Butler wrote, in the language and perspectives of the times,

> These are women of recent immigration, raw from their peasant earth, unacquainted with the language and ignorant of the ways of this country. They live in colonies of their own people. They accept the work and the conditions that go with it which more often than not are unnecessarily degrading. Except where the nervous speed of American girls is in demand, these foreign women who ask neither for comfort nor for cleanliness nor higher wages, form the group characteristic of the trade.

It was the volume *Work-Accidents and the Law*, though, that spelled out in numeric detail and for the first time the causes and impacts of injuries and deaths in steel, zinc, and other metallurgical industries in Pittsburgh and, by extension, in Donora, Homestead, Monessen, and other industrialized towns as well. The work-accident volume was based on studies conducted by brand-new law school graduate Crystal Eastman.

Born June 25, 1881, in Elmira, New York, Catherine Crystal Eastman was a tall, impressive beauty with dark wavy hair, a square jaw, and a graceful neck. She had kind, soft eyes that belied an inner passion for equality, internal dynamism, and a penchant for direct action. She called herself a "militant idealist." Freda Kirchwey, longtime editor of *The Nation*, said in Eastman's obituary, "When she spoke to people—whether it was to a small committee or a swarming crowd—hearts beat faster and nerves tightened as she talked."

Eastman started her stellar career right after college. She graduated from Vassar College in 1903, finished second in her class at New York University School of Law in 1907, and immediately joined the Pittsburgh Survey team. Eastman's groundbreaking work with the survey propelled her into a position on a just-formed and highly prominent labor panel in New York, with the comically long name of the New York State Commission of Employer's Liability and Causes of Industrial Accidents, Unemployment, and Lack of Farm Labor, later renamed the Wainwright Commission, after Jonathan M. Wainwright, a prominent New York attorney and politician.

As a founding commission member Eastman authored the state's first worker's compensation law. That law became the model for similar laws in many other states. Eastman was also the "founding mother" of the American Civil Liberties Union, organized in 1920 to protect the constitutional rights of all Americans. By then a nationally recognized lawyer and activist, Eastman went on to coauthor, with women's rights activist Alice Paul and others, the first equal rights amendment in 1923. The Lucretia Mott Amendment, named for the renowned abolitionist, stated, "Men and women shall have equal rights throughout the United States and every place subject to its jurisdiction. Congress shall have power to enforce this article by appropriate legislation." The Mott amendment was proposed in every congressional session from 1923 to 1943 and, shamefully, was rejected every time.

Eastman succumbed at forty-seven to a chronic kidney condition, caused by a severe bout of scarlet fever as a child. Even though she died before her time, her seminal study, *Work-Accidents and the Law*, saved the lives of workers everywhere.

Social activists, engineers, and politicians used the data presented in Eastman's remarkable study to support a potpourri of reforms, including the abolishment of child labor, improvements in public health, and the development of more refined civic planning. Eastman provided illuminating statistics to guide policymakers in the future, all the while focusing on the workers themselves in a clear, compelling writing style.

For instance, she detailed the causes of death of the 195 men killed in Pittsburgh that year. She determined that the most dangerous machines to work with were not the blast furnaces, though they were certainly dangerous, nor the massive rolling machines, nor the giant ladles filled with molten steel, nor the elephantine contraptions that sheared thick steel plates to precise sizes. The most dangerous machines by far were trains and cranes.

Nearly 40 percent of all fatalities at steel mills that year occurred during the operation of cranes and of narrow- and broad-gauge railroads. Eastman told of four Slavic men killed in a horrifying crane accident, when a ladle carrying molten steel fell on them. Two men were instantly consumed by metal. One of the others was crushed. He was just eighteen years old and had been in the United States only three months. The lad didn't want to work in any area below a ladle, but his supervisor had insisted. Wrote Eastman, "A few minutes later a hook broke and the bucket, weighing with its load 6,900 pounds, fell and crushed him. The crane operator testified at the inquest that he 'never knew of hooks being inspected.'"

Railroads were especially dangerous. People working with or around the large, broad-gauge trains or, especially, trying to jump on them during movement, faced significant risk of injury or death. Eastman determined that eighteen Pittsburgh workers were killed in broad-gauge railroad accidents during the 1907–8 study period. Four men were run over, two caught between cars, two killed in train crashes, and four killed trying to jump onto moving trains.

Narrow-gauge, or dinkey, trains could be dangerous as well, especially for the engineers and brakemen operating them. Dinkey trains frequently jumped the track, often doing so without warning. One engineer was killed when the pin connection between his engine and the train's first

FIG. 13.2. Young worker at American Steel & Wire in Donora. Courtesy of the University of Pittsburgh Bruce Dreisbach–American Steel & Wire Company Photographs.

car suddenly broke. The engine burst ahead and jumped the track, crashing into some broad-gauge cars nearby and killing the operator. A total of thirteen men were killed in dinkey train accidents that year.

Eastman's report didn't just tabulate data and tell stories. It also offered concrete suggestions for improving worker safety. To reduce train accidents, for instance, she suggested that

overhead or underground passages might be built where men must cross tracks. Where this is impossible, trains might be run more slowly. Rules forbidding mill employees to ride on the trains might be more rigidly enforced. Ignorant non-English-speaking foreigners employed on or about tracks and cars might be under direction of a foreman who speaks their tongue. Lights and signals are especially important in mill yards, because the regular noises of the mill soon dull the ear to the sound of bell or whistle.

Eastman, like other Pittsburgh Survey authors, offered commonsense suggestions without a hint of blame. Unfortunately, for all of its research, its tragic stories, compelling data, and reasonable suggestions, the survey was largely a bust. According to Meryl Nadel, professor of social work at Iona College in New Rochelle, New York, "Pittsburgh's business and professional elites were receptive to some recommendations, gave lip service to others, and rejected those they found inconsistent with their perspective."

Pittsburgh was one of the most industrialized cities at the time and home to such behemoths as Alcoa, Bayer, Heinz, US Steel, and Westinghouse. US Steel in particular seemed ubiquitous in and around Pittsburgh, its mills lining both sides of the Monongahela. In many ways US Steel *was* Pittsburgh. As the survey's leader Kellogg put it, "The largest employer of steel workers in Pittsburgh is the largest employer of steel workers in the country as a whole; and the largest employer of labor in America today. That employer is in the saddle."

Riding that saddle was the Pittsburgh triumvirate of Henry Clay Frick and Andrew and Richard Mellon. The city's powerbrokers almost certainly read the Pittsburgh Survey or were at least briefed on it. Eastman's chapter on work accidents and the one on steel workers, *The Steel Workers*, by John. A. Fitch, probably received the most scrutiny. Nadel wrote that "Kellogg and the other investigators naively expected their conclusions to be greeted with popular acceptance followed by legislative and other governmental reforms."

It was not to be. Reactions to the report came swiftly and sharply. The survey pointed out management failures and motivations of profit over safety, concepts that flew in the freshly manicured face of Pittsburgh's most elite residents. Mellon, Frick, Carnegie, and the like had worked diligently to portray the city as a bastion of success, a preeminent hub of industry and innovation, the nation's quality of life polestar. The Pittsburgh Survey undercut those ideals and dealt a heavy blow to the city's national image.

Kellogg and his investigators were accused of being "outsiders," which they were; none of the main investigators were Pittsburgh natives, though some of the door-to-door surveyors hailed from the city. The report was criticized as "neither evenhanded nor thorough." Bookstores refused to sell the reports in book form after they had read installments published by the *Post*. Pittsburgh newspapers began to refuse to publish articles from the survey team. Whether they were pressured to do so by outside forces or did so on their own, the results were the same.

Never mind that the surveyors had collected an extraordinary amount of qualitative and quantitative data, that they had provided the city and its leaders with more information and insight into the community than ever before, and that they had suggested specific, not terribly expensive corrections for reducing injuries and saving the lives of untold numbers of workers. Had the report succeeded in motivating US Steel, for example, to make even some of the recommended safety improvements, changes would probably also have been made to factories in Donora, Clairton, Homestead, and other metal plants throughout the state. But precious few changes occurred.

In the end it seems clear that the young surveyors and their leaders had greatly underestimated the will of Pittsburgh's industrial elite and overestimated the elite's interests in much of anything else but profit. Without the backing of one or both of the Mellons, or Frick, or even the aging Carnegie, who by that time was focusing almost exclusively on philanthropy, the reports were bound to land in the slag heap of well-intentioned but ultimately unsuccessful ventures. Industry called the shots back then, and with US Steel the largest employer in Pittsburgh, the largest steelmaker in the country, and the largest corporation in the world, anything that ran contrary to its interests would simply not be tolerated.

Enter a young steelworker named Andrew Posey.

14

THE PERSISTENT
LEGEND
OF YOUNG
ANDREW POSEY

OF THE THOUSANDS OF ACCIDENTS THAT OCCURRED AT DONORA FAC-
tories over the years, none have captured the imagination of Donorans
near and far quite like Andrew Posey's. Posey, an Independence Day baby
born in 1897, enlisted as a private in the US Army Ranger program on
October 15, 1918. He was assigned to an army training camp in Pittsburgh.

Germany would sign an armistice to end World War II in less than a
month in a railroad dining car in the Forest of Compiègne, France. With
no war now to fight, Posey was demobilized and honorably discharged on
December 9, 1918, having served not even three months and having never
seen duty outside Pennsylvania.

After his discharge Posey found work in Donora as a ladle stopper in
American Steel & Wire's open hearth furnace. Ladle stoppers were responsible for making sure the nozzle, or stopper, at the bottom of the ladle was
clear before a furnace's heat could be poured in. To clear a stopper the
worker would climb into the ladle itself—a perilous task considering that
a furnace holding thousands of pounds of three-thousand-degree mol-

ten steel was directly overhead. After the stopper was cleared, the worker would climb out and signal workers above that it was "all clear to pour." The molten steel would then pour out of the furnace, sparks of white-hot metal flying in all directions. Spectacular, but potentially deadly.

On Thursday, January 8, 1920, a little over a year after being discharged from the army, Posey was inside an empty ladle, adjusting the stopper, when suddenly molten steel poured into the ladle. The steel instantly vaporized Andrew, the molecules making up his body becoming part and parcel of the very steel that killed him.

Mill whistles blasted to announce that a serious accident had occurred. Posey's fellow workers rushed to the ladle to see if they could help or perhaps just to witness the scene of a violent death. The county coroner, William Greenlee Sr., listed as the cause of death for the twenty-two-year-old as "Burned while at work in Open Hearth Dept. of A. S. & W. Co. Donora, Pa. Body entirely consumed by hot metal. Accidental."

Two days later a simple thirty-four-word article appeared in the *Pittsburgh Post-Gazette*. So common were steel mill injuries and deaths that the article was relegated to the bottom of page 7, adjacent to an equally simple article called "Prairie Itch Epidemic in Sharon." The Posey article read, in its entirety:

HOT METAL SHOWER KILLS YOUTH

DONORA, PA., Jan. 9—(Special.)—

Andrew Posey, aged 24, employed in the American Steel and Wire Company plant, was burned to death yesterday when hot metal came down upon him while he was adjusting a stopper in a ladle.

Such a small notice for such a horrific accident. An inquest was apparently held by Washington County Deputy Coroner C. E. White the following Monday, January 12, but the report has not survived.

Stories about the young man's death spread rapidly. At some point the basic story began to morph, eventually taking on practically mythic proportions. In many stories told about Posey's death, an explosion started the whole thing. According to a 2006 letter written by Andrew's sister

Margaret, who was two years old at the time of her brother's death, the furnace above Posey's head exploded. Ore that had been dumped into the furnace was said to be frozen, and the frozen ore triggered an explosion.

Open hearth furnaces sometimes did explode, but not from frozen ore. Chris Pistorius, professor of materials science and engineering at Carnegie Mellon University, explained why the explosion theory doesn't add up. First, after each heat of steel the furnace would have been "tapped dry," meaning all molten metal and slag would have been removed from the base of the furnace. No liquids would remain.

Workers would then have set down layers of limestone and bits of used steel, called scrap. Hot metal, consisting of liquid iron from one of the blast furnaces, would then have been poured into the furnace. Pistorius said that sometimes a "vigorous boiling" occurred when charging the hot metal, which could have caused slag to foam out of the furnace and, potentially, burn anyone working nearby. "While it does not seem likely that ice in scrap would have caused explosions during charging of the scrap," concluded Pistorius, "there certainly was the possibility of an ejection of foaming slag from the furnace during the subsequent hot-metal charge."

If Posey had been killed by foaming slag, though, contemporaneous descriptions almost certainly would have differed in their wording. No mention was made in those accounts of any kind of explosion, regardless of origin. If an explosion did in fact occur, it seems doubtful that any reputable reporter would have kept that information secret. It is possible, though unlikely, that American Steel & Wire officials pressured reporters who showed up at the mill to keep that part of the story quiet. But if that were the case the entirety of the incident would have been squashed. No, there was most likely no explosion.

Another version of the accident has Posey falling into the ladle. If that story is accurate, and it might be, the most likely scenario would have him falling *back* into the ladle after he had already adjusted the stopper. He might have slipped as he crawled out of the ladle, and the furnaceman above might not have realized the situation and dumped his load of steel too soon.

Suicide has also been mentioned as a possible reason for the accident. That theory is based on the concept that as a serviceman during World War I Posey could have suffered from shell shock, a condition known today as post-traumatic stress disorder. Posey never saw active duty, how-

FIG. 14.1. Two workers stand on a platform at the Homestead Steel Works to adjust the stopper of a ladle as molten metal pours from the bottom. Courtesy of the William J. Gaughan Collection of the University of Pittsburgh.

ever, so there is little likelihood that he suffered such a disorder—at least not from his time in the service.

Still another story has Posey signaling the furnaceman to release the steel and then suddenly jumping back into the ladle to fix one last thing before jumping out again. If Posey had a bit of tempestuousness in him, a bit of invincibility, this explanation holds some merit. He was twenty-two, three years younger than the average age at which the rational part of the brain fully matures, so he hadn't yet fully matured in his thoughts or reactions. However, Posey would have known how quickly the furnace-men could have dumped molten steel into the ladle, so to assume that he believed he had enough time to climb into the ladle, fix something, and then climb back out before the steel poured all over him seems beyond the pale.

Other stories circulated as well, including one claiming that Andrew's parents, Andrew and Susie, along with their nine surviving children, demanded that the mill bury the young man's remains. There were no remains, of course, just the steel that had poured into the ladle. As the story is told, mill officials at last acquiesced to the family's demands and buried the "remains." Whether those remains consisted of the now-solid heat or the entire ladle with the heat inside varies with the person telling the story.

Visitors to Donora may be struck by the number and variety of stories about Posey's death that persist today. Studies into how the human brain forms memories indicate that a single event typically gives rise to a variety of perspectives of the event. Each of the mill workers who were standing near the ladle when Posey died, if any actually were, would have remembered the event somewhat differently. Alin Coman, associate professor of psychology and public affairs at Princeton University, explained that "people have different perspectives when they perceive an event. These perspectives are shaped by their motivations, beliefs, and values. The interaction of those psychological constructs can drastically affect the way one sees the world."

In addition, when workers near Posey began telling others about the event, their own descriptions would have begun to shift. The witnesses would have, consciously or subconsciously, emphasized or deemphasized different aspects of the event, depending on their listeners. "People have different positions in their social networks," said Coman, "which means that they might have access to different pieces of information."

The storyteller may share certain information with some listeners but not others. The listeners in turn remember different features of the story, depending in part on whether they felt more or less trust in the story-teller. Either way the level of trust can spark biases that affect the story told in subsequent conversations.

So how did Andrew Posey actually die?

One possible reason: miscommunication. Miscommunications led to a great many injuries and deaths in steel mills. Eastman, in her study of workplace accidents, discovered that most steel mill accidents resulted not from faulty equipment but from simple, understandable human errors. In one incident, typical of miscommunication events, a pipe fitter had descended into a pit beneath a roller, a machine that transforms red-hot blocks of steel into bars, beams, or rails. The man controlling the roller, unaware that the pipe fitter had entered the pit, signaled to another worker to raise the roll table, which operates by weights moving up and down in a pit. "As the table was raised," said Eastman, "one of these weights struck and killed the pipe fitter." The same type of miscommunication might well have resulted in Posey's death.

Perhaps the furnaceman wasn't paying close attention to his controls. He might have been distracted and released the molten steel too soon.

Or perhaps Andrew did indeed jump or fall back into the ladle, for whatever reason, just as the steel was released.

Historians at the Donora Historical Society generally consider the incident more mystery than fact. Andrew Posey was killed when he was consumed by molten steel poured from a furnace above the ladle he was working in. Those are the facts; anything beyond them is pure speculation.

As for the heat containing Posey's remains, there is no plausible evidence to support the theory that the heat or ladle were buried somewhere. To think that any official of American Steel & Wire or its owner, US Steel, would have done anything to memorialize one of its workers is to vastly overestimate the level of concern management expressed about them.

A somewhat similar accident occurred at a steel mill in Pittsburgh. A fledgling steelworker, Charles Walker, described in his diary a story about an accident at a Bessemer furnace, an older type of furnace similar to an open hearth. Apparently the plug for a tap hole at the bottom of a furnace had broken and let loose a stream of molten metal. "It caught twenty-four men in the flow—killed and buried them," wrote Walker. "The company, with a sense of the proprieties, waited until the families of the men moved [away] before putting the scrap, which contained them, back into the furnace for remelting."

It seems likely that the heat containing Posey would have suffered a similar fate. Mill workers at the time were viewed as essentially chattel. If one worker was injured or killed on the job, another would be assigned to take his place. Reliable accounts abound of workers who had suffered burns, sprains, mild concussions, or other non-life-threatening injuries being expected by management to return to work immediately. Workers with more severe injuries might have been allowed more time off, but not much. Workers who died, well, the mill had to keep running. Clean up the mess, and move on.

Unfortunately there are no known diaries or other personal accountings that describe the Posey accident. Lynne Page Snyder, whose 1994 doctoral dissertation examined the 1948 fog in detail, described the difficulty she experienced in finding contemporaneous, resident-written accounts of the smog or anything related to the mills that might paint the town in a poor light:

Area residents, weary of decades' worth of interviews, were often unwilling to speak at all. When they did, their presentation was usually a set piece, shaped by subsequent exposure to environmentalism and polished through years of practice in the giving of interviews. Key primary materials had been given away years ago to researchers of one ilk or another, or disposed of with a vengeance when encountered during the clearing of attics and basements. Donorans, it seems, have tried repeatedly to purge the bitterness of memories about the smog and its legacies for them.

What actually happened to Posey and his heat may never be fully known. Accident records at American Steel & Wire have never been made public and in fact might not exist. It isn't hard to imagine company leadership locking away whatever notes might have been recorded about the incident, nor is it hard to imagine those records having been destroyed at some point, innocently or purposefully. Eastman pulled no punches in deriding steel companies for their secrecy in handling worker accidents:

One does not need the training of an engineer nor the experience of a mechanic to see how the side of a steel car may slip from crane chains and fall on a man below, or how a machinist walking a greasy beam 60 feet in the air may lose his balance, or how a pile of iron pipe may fall and crush a man or two if something gives out at the bottom of the pile. The large steel companies have long defended a policy of silence with regard to the number and character of their accidents, largely on the ground that any other policy would result in unintelligent hysterical outcry and clamor on the part of the public. If accidents in steel mills were altogether a result of processes which only experts can understand, there might be some reason in such a policy. But finding that at least fifty-seven per cent of the fatal accidents studied were due to ordinary understandable causes, we can maintain not only that the public has a right to know the facts but that its possession of this knowledge is an important factor in the prevention of accidents.

At some point in the latter part of the 1900s Posey's brother Joseph constructed a gravesite in an open area of the mill complex thought to house Andrew's heat or ladle. But no one really knew where, or even if, the heat or ladle had ever been buried.

In 1995 a local group called the Mon Valley Progress Council decided to determine once and for all whether the site contained any large steel object. It hired an engineer to have the site excavated. The engineer found nothing but dirt, rocks, and slag remnants. "We conclude," wrote Samuel M. Bitonti in his report to the council, "that this site is a memorial only and [that] no slab or ingot of steel is located within the area of our investigation."

Stories about Posey's accident have never strayed far from the collective consciousness of the Donora community and are routinely told to visitors today. Thousands of visitors over the years visited Posey's memorial. Perhaps they wondered what the young man must have felt when, or if, he realized his life was about to end. Perhaps they wondered what life must have been like for mill workers, how dangerous it must have been, and how tough those workers must have been to endure such tragedy. Perhaps they thought of the men and women who toiled in metal mills elsewhere and who died or were disfigured as a result.

No matter how Andrew died or what happened to the heat he perished in, his gravesite and the stories surrounding his death continue to serve as quiet, contemplative memorials to steelworkers everywhere.

15

DEATH ON THE RIVER MEUSE

FOG ROLLED INTO HUY, ONE OF MANY SMALL TOWNS ALONG THE Meuse River valley in Belgium, on December 1, 1930, a Monday. Autumn and early winter fogs were common in that part of the valley. Steep hills on either side of the river, like the windows in a smoker's car, tended to restrict fog to the basin. Townspeople in Huy (pronounced like the French *oui*) ambled by Li Bassinia, a fountain in the middle of the town plaza, a favored gathering place.

The fountain, built in 1406, features statuettes of three saints (Catherine, Domitian, and Mengold), the last Count of Huy (Ansfrid), and a town watchman who would stand in a church belfry and blow his horn to alert townspeople that an enemy was approaching. The entire fountain is surrounded by an ornamental wrought-iron canopy topped by a bronze, two-headed eagle. Townspeople gathering around the fountain that day ate, drank, talked, and people-watched, just as they had done for centuries.

The fog seemed not at all different from other fogs. The mist dulled the typically golden hues of the intricate Bethlehem Gate that guarded the entrance to the Notre-Dame de Huy cathedral. The façade sculpted on the gate depicts three biblical tales: the nativity of Jesus and annunciation to the shepherds on the left, the massacre of the innocents on the right, and the adoration of the magi featured prominently in the center. That day the façade seemed more brown than gold.

People went about their daily lives and started their work week as they always had. When they returned home for the night, the fog remained. The fog thickened the next day, and again the next. Fog in the valley generally didn't last longer than a day or two, but to residents it was just another fog.

By the third foggy day in Huy some of the more fearful residents began to think the fog would bring about the end of time. Residents who walked through the square on Wednesday could barely make out Li Bassinia until they were almost upon it, and many of them coughed as they walked quickly along. By Thursday hundreds of people in town were coughing, choking, and having trouble breathing. Something was terribly wrong, and not just in Huy but all along the valley northward to the city of Liège, one of Belgium's cultural hubs. Between Monday and Friday, when the fog finally lifted, thousands of valley residents were sickened and sixty were dead. Animal carcasses lay sprawled about in the fields, "like flies sprayed with poison gas."

Not quite two decades later Donora would suffer an identical fate.

The deaths along the Meuse River shocked not only Belgium but nations around the world. The *Manchester Guardian* in England published a story on December 6 with the headline, "Deadly Fog in Belgium. Poison Gas or Factory Fumes?" Another British paper, *The Observer*, not only noted the tragedy but also decided, after just two days, on the cause of the deaths, choosing as the headline for its Sunday edition, "60 Fog Deaths in Belgium. Natural Causes Responsible. Glacial Mist and Weak Lungs."

US newspapers from Boston to Shreveport to Sacramento picked up the story, generally from either the United Press or Associated Press news agencies. The *Daily Republican*, Monongahela's main newspaper, published a United Press story on December 6 titled, "Death-Bringing Fog Lifting in Belgium: Dampness Alone Believed to Have Caused or Hastened 42 Deaths in Meuse Valley—Memories of Poison Gas Revived."

The *Pittsburgh Post-Gazette* the same day went all-in on the poison-gas theory. Its headline read, "Mystery Poison Gas Kills 64 in Belgium."

Nobody knew what had really happened. Theories abounded, including that the deaths were caused by a resurgence of the so-called Spanish flu, a volcanic eruption, toxins released from stored chemical weapons left over from World War I, or infectious microbes carried on the wind from, of all places, the Sahara Desert. Even J. B. S. Haldane, a well-known British geneticist and scientist, entered the fray, initially claiming that the problem was "something like the Black Death," referring to the bubonic plague of medieval Europe. "I don't believe the epidemic could have been caused by war gas," Haldane explained. "They have been having floods in that district lately, and it is possible that may have something to do with it."

None of the theories proved remotely accurate. The tragedy in the Meuse Valley, as the eminent Dr. Haldane would soon concede, was caused by a common, cool-weather event in combination with emissions from steel and zinc plants lining the river—the same deadly combination that would strike Donora eighteen years later. Would Donora heed Belgium's warning?

The Meuse River is one of two main rivers in Belgium, the other being the Scheldt. The Meuse flows northward, like the Monongahela in Pennsylvania. It begins its journey as a shallow gulley in an area called Pouilly-en-Bassigny, about 160 miles southeast of Paris. Bronze plaques on a simple commemorative stone marker show a map of the river on the top, and on the bottom the words "Meuse endormeuse et douce a mon enfance, qui demeures aux pres ou tu coules tout bas," or "Sleep casting Meuse, sweetening my childhood, meandering the meadows, where you run lower and lower."

From that damp little gully the Meuse grows in width and depth and wiggles northward into Belgium. The river takes a strong westward turn at a town called Namur, then waggles in a long, lazy west–east bend to its final destination, the North Sea. The tortuous Meuse flows through Belgium in a narrow, deeply cut valley. The valley walls between Huy and Liège are particularly steep, about 450 feet high in Liège, Belgium's third-largest city.

Huy and Liège were industrial centers of the valley, with tiny Engis sitting halfway between them. Engis was home to the Métallurgique de Prayon, or Prayon Metallurgical, the largest zinc manufacturer in Belgium when it opened in 1882. Just twenty miles east of Prayon lay the Plombières-Altenberg-Moresnet mining district, rich in zinc and coal. Prayon was then just one of eight zinc smelters in the fifteen-mile stretch between Huy and Liège. That section of the river valley also contained six coke mills, five steel mills, and two fertilizer plants, along with an array of quarries and small factories.

Belgium's economy in 1930 had not yet suffered as much from the Great Depression as it would a year later. In fact the economy was robust enough for the nation to hold celebrations in three cities for the one-hundredth anniversary of its independence from Dutch rule. Antwerp put on an industrial and commercial exposition, Liège put on an arts, crafts, and electricity exposition, and Mons showcased famous artwork, as did the nation's capital, Brussels.

Brussels also held a centenary parade that ran from the Palace of Justice and along Regency Street before taking a slow left turn into Place Royale, a huge plaza built in the style of Louis XVI. The parade featured columns of marchers waving Belgian flags, gaily dressed trumpeters on horseback, and floats of all kinds, including one of a Trojan horse more than twenty feet high, with four riders seated in the horse's saddle. They all passed in front of Belgium's beloved king and queen, Albert I and Elisabeth of Bavaria.

A few weeks later, on the first morning of December, Belgians woke up to a fog that covered much of Western Europe. Fog stems from a meteorological event called a temperature inversion, sometimes called an anticyclone. In a temperature inversion a layer of warm air sits on top of cooler air at earth's surface. Under normal day-to-day conditions, the higher the elevation, the cooler the air, as any mountain climber can attest. In the morning the sun warms earth's surface, which heats the ground and the cooler air just above it. The warm air then rises into the atmosphere, where it cools. At the end of the day, as the sun goes down, the ground and the air above it cool again.

In warm weather the ground temperature doesn't differ much from the air temperature. In cooler months, though, the difference between the two temperatures can vary widely. Air at any temperature can hold moisture in the form of gaseous water, or water vapor, but cooler air can hold much more water vapor than can warm air. Nights during late summer and fall can become quite cool in certain areas and are coolest shortly before sunrise. If air near the ground cools enough it can become saturated with water vapor, a point meteorologists call the dew point, when fog forms.

"As the air right above the ground is cooled down, it becomes stable," said Neil Donahue, professor and director of the Steinbrenner Institute for Environmental Education and Research at Carnegie Mellon University in Pittsburgh. "The air wants to stay there because it's cold and dense. Cold air will tend to sink, and warm air will tend to rise up. Whenever you have a situation where the ground is getting cooled off, air will tend to pool. Cold air will tend to form little puddles anywhere there's a dip."

Fogs that form that way are known as radiation fogs and tend to be thicker in areas such as river valleys. Streams and rivers in a valley serve as a concentrated bank of water that can evaporate, the rate changing along a river mainly according to the temperature of the air and turbulence of the water. When only slight breezes flow through a foggy valley, the fog can linger longer than would the same kind of fog over land.

Radiation fogs tend to stay in one place anyway, valley or not. After the sun appears and warms the air at earth's surface, radiation fogs tend to "blow off." Donahue explained that "sunlight is not absorbed by the fog; it bounces around. The fog is white, but the sunlight is absorbed by the ground. The sunlight starts to heat the ground, and that reverses the process. The air warms and turns the liquid fog droplets back into vapor. Eventually the fog dissipates. That's the normal condition."

What happened in the Meuse Valley was not at all normal. Air in the valley was still, with barely the whisper of wind anywhere along the river. Ground-level air remained cool, with a layer of stagnant, warmer air lying about 250 feet above the valley floor. People looking down from the surrounding cliffs, some 350–450 feet high, would have seen below a valley filled with soft, fluffy clouds, but they wouldn't have seen the valley floor.

People leaving their homes that morning would have been met with the same kind of fog they had experienced many times before. As the day progressed the fog would have become a bit darker, a bit thicker, a bit more irritating to the eyes and throat. Each day thereafter the fog would have become even darker, even thicker, and even more irritating, until finally burning eyes and a dry, hacking cough became the order of the day.

Residents might not have thought much at the time about the coke, steel, and zinc factories lining the valley, mills that continued to pour soot, dust, fumes, and gases out of their chimneys into the air. None of the chimneys were taller than about two hundred feet, about fifty feet short of the bottom of the warm-air canopy. All of the fumes and gases pouring out of those chimneys, all of the particles of soot and dust, all of those pollutants discharged into the air bumped up against the canopy and hung there. The longer the fog lasted, with its canopy holding down the smoke, the more toxins permeated the air people breathed.

The temperature inversion that started Monday along the Meuse hung around all week, not letting up until Friday, when winds from the east broke up the fog and cleared the air. The valley had not experienced fog lasting more than three days for eleven years. In fact, until 1930 fog had lasted that long only four times since the turn of the century: in 1901, 1911, 1917, and 1919. The fog of December 1930 was not only one of the longest in Belgium's history; it was also the deadliest.

The fog reached its thickest on Wednesday, a day that also saw a sudden increase in the number of residents experiencing eye and throat irritation, wheezing, chest pain, shortness of breath, and spasms of violent coughing. The sickest people developed rapid breathing, frothy sputum, and mottled blue skin—all signs of pulmonary edema, a life-threatening condition. Without treatment pulmonary edema is fatal. Victims suffer a terrifying struggle for breath that devolves into a kind of slow suffocation. So little oxygen reaches the brain that the person falls unconscious. Breathing slows, arms and legs turn cold, and at the last, a final, faint breath escapes the lips.

Many residents that week pulled their World War I gas masks out of storage and began wearing them in the hopes of avoiding whatever was in the air. Physicians throughout the valley worked day and night to help people keep breathing. Hour after hour the doctors saw patient after patient having coughing fits and choking spells, conditions they hadn't seen since the Great War, when Germans stunned Allied forces with chlorine gas in 1915 and mustard gas in 1917.

The mayor of Engis, an industrial town south of Liège, admitted being bewildered by the entire ordeal. "The 3,000 inhabitants of my little town are terror-stricken," he said. "Besides the dead, there are hundreds suffering with a strange disease." The mayor spoke later with the *Sunday Mirror* in London and described his anguish:

Hundreds are ill, including myself. I have risen from a sickbed, to speak to you. The majority of these dead are middle aged and elderly, but there are also girls of twenty. Some of the victims died within two hours. An impenetrable pall of blackness has been around us since Tuesday. We have closed all doors and windows, and stopped every chink with paper and rags. The streets are deserted corridors, inky with blackness. The fog is so dense that one cannot see three yards away. I cannot express any opinion as to the cause of the deaths, but it is widely believed that they are due to chemical fumes, from factories, which the fog has prevented from dispersing.

Sixty people died during the fog, mostly those who were elderly and already suffering from asthma, heart conditions, or other conditions that left them unable to withstand the stress brought on by the toxins. The greatest number of deaths occurred in little Engis, where fourteen people perished over the two worst days of the fog. By comparison, Engis would normally have had sixty-five deaths over an entire *year*.

In the immediate aftermath of the tragedy numerous health officials weighed in with their own versions of the cause. An early report submitted by Dr. Lacombe from the Liège public health administration and Dr. Timbal from the Ministry of the Interior's public health administration claimed that a sudden temperature drop and "the fog alone" explained the deaths. Why the report left the town's zinc smelter and other heavy industries out of the equation was never revealed. Lacombe and Timbal released their surely comprehensive report on Saturday, December 6, a single day after the fog lifted.

A noted French physiologist, Jules Amar, stood by his own illogical assessment. Amar, author of a 1920 landmark book in physiology called *The Human Motor: Or, the Scientific Foundations of Labour and Industry*, believed that the deaths had been caused by the inhalation of so many water droplets in the air that people essentially drowned, a condition he termed "poisoning by humidity."

Common sense soon prevailed. Pierre Nolf, the king's physician and head of Belgium's Red Cross, put into words what calmer heads might have been thinking all along: "Pure fog has never killed anybody." The Belgian government agreed and formed two committees, the main one being a "multidisciplinary judicial commission" to "determine the mechanism of the accidents, deadly or not, which occurred in the valley on the fourth and fifth of December." Forty-year-old physician Jean Firket was placed in charge of the judicial investigation. Firket taught pathology at the University of Liège and was already highly regarded in the field of forensic medicine. Besides Firket the commission consisted of a toxicologist, meteorologist, veterinarian, two chemists, and one more physician.

The team conducted ten autopsies on humans and several on cattle. The human postmortems consistently showed alterations in lung tissue; no other organs were damaged. Blood samples contained evidence of the toxins zinc, sulfur, and arsenic. Examinations of deceased cattle indicated that they had suffered from two lung conditions, both of which impaired the exchange of oxygen and carbon dioxide.

The team also tested air in the valley for known pollutants. They found that the factories were releasing a total of thirty substances, including carbon dioxide, carbon monoxide, hydrogen sulfide, arsine (an arsenic compound), acetylene, and a variety of carbon and sulfur compounds. After testing the concentrations of the compounds and considering each substance's toxicity, the team concluded that sulfur dioxide, sulfur trioxide, and sulfuric acid, carried on tiny dust and soot particles, were present in high enough concentrations in the air to cause severe toxicity and death.

Firket and his team concluded that "the sulphur produced by coal burning has had a deleterious effect, either as sulphurous anhydride or acid, or as sulphuric acid; the production of which was made possible by unusual weather conditions." As to the actual source of those toxins, Firket blamed only "coal burning by industry and by private households." Again the mills along the river escaped direct blame, a pattern that would later repeat itself in Donora and again, at least for a time, in London, when in 1952 that city experienced twelve thousand deaths in what is known now as the Great Smog of London.

The Meuse commissioners suggested a number of preventive measures for avoiding similar tragedies, including the construction of a weather observatory in the valley that could alert mayors in susceptible towns to prepare for an impending temperature inversion. The mayors could then distribute gas masks to residents and transport more vulnerable citizens to higher ground, steps that would head off some deaths but did nothing to address what the commissioners believed was the root cause: toxic emissions from the mills.

Mill owners refused any assignment of blame and argued that they simply could not "stop or shelve the ovens for a period of a brouillard." Clearly no fog was worth any interruption in business as usual. Community leaders along the valley fell in line, feeling rather disinclined to blame the industries that were keeping their towns afloat. Perhaps most telling, the Belgian government considered the sad situation a normal, almost an expected part of a growing, industrialized society. In the end, the "official" cause turned out to be the fog itself, a force majeure, and not the industries suffusing the fog with poisons. A small, inconspicuous statue now stands in Engis, a testament to the victims of the smog. The statue depicts a young woman named Louise. She is kneeling, her long hair falling over her left shoulder, her head tipped to the right, her clasped hands protecting her face from some unknown assailant. The plaque on her base reads:

> Louise was pretty,
> Louise was 20,
> She was coming back from the ball,
> She was a child . . .

In memory of the sixty dead, young and old, of Amay, Engis, Flémalle, Seraing, victims of the atmospheric accident of December 1930 in the greater Liège region.

Any human enterprise, even an industrial one, is capable of improvement.

16

DECISIONS,
DECISIONS

AT THE TIME OF THE MEUSE DISASTER A FARMER NAMED GEORGE GLIWA had a lawsuit in progress against US Steel. Gliwa lived in Lincoln, Pennsylvania, about eight miles north of Webster and directly across the river from the Clairton Works, a large US Steel–owned coke plant. He had for years suffered the ill effects of smoke pouring from the plant's chimneys. Initially Gliwa's suit centered on a US Steel subsidiary, Carnegie Natural Gas, trying at the time to obtain a right-of-way through Gliwa's seventy-one-acre farm. Gliwa refused to allow it. The gas company sued for access.

Gliwa's attorney, Joseph Pinkasiewicz, countersued. Gliwa had told Pinkasiewicz that his wife's health was poor and that his cattle and farmland had been suffering from the "sticky tar" effluents emanating from the coke works. "I called the state farm bureau to find what was causing my cattle to die," said Gliwa. "After examining the place they said it was the acid in the air from the mill smoke. I can't even feed the cattle hay grown on my farm. The bureau told me to buy my hay and stop up the cracks in my barn to keep the acid out." He also said that after harvesting four thou-

FIG. 16.1. Beehive ovens, like these inoperative ovens in Mount Pleasant, Pennsylvania, were often built into the side of mountains. Carl Mydans, Farm Security Administration, Office of War Information Photograph Collection, Library of Congress, 1936.

sand cabbage plants there was barely enough healthy cabbage "to make half a barrel of sauerkraut."

Pinkasiewicz added claims for damages to the countersuit, citing Agnes Gliwa's poor health and numerous damages to the couple's crops and farm animals. Lawyers for US Steel believed they could easily outlast the poor farmer, so they hammered Pinkasiewicz with procedural motions, one after another, until both suits were stayed indefinitely, in November 1930.

News of the Meuse Valley incident, however, might have motivated the persistent Pinkasiewicz to push even harder on a strategy he had been working on to slay the mighty Clairton Works dragon.

Built in 1901 by St. Clair Steel Company, the Clairton Coke Works was purchased in 1904 by US Steel. Coke ovens remove sulfur and other impurities from soft coal, leaving behind nearly pure carbon. Processes used at the new coke mill were considered far superior to the beehive oven processes they replaced. Invented in 1849 beehive ovens dotted the national landscape in the early 1900s. A total of 100,362 individual ovens across the country were being managed by 548 coke plants. The ovens burned coal day and night, every day, all year round.

By contrast the new coke mill provided a hundredfold increase in coke-producing capacity. Ovens in the mill were also able to trap much of the gas that beehive ovens would have released into the atmosphere and reuse it to produce coal tar for the manufacture of paint dyes and numerous other industrial products. The reused gases, then, provided the mill with an additional source of revenue.

Liquid wastes from the mill were another story. Officials at Clairton Works wanted to release liquid wastes, which had no further industrial uses, into the Monongahela River. The nearby town of McKeesport objected. The town was granted an injunction against the dumping because it drew its water from locations directly downstream from the mill. So US Steel decided instead to implement a process that would aerosolize the wastes and pass the mist into the atmosphere. Locals quickly found that those aerosols contained ammonia, making the air sticky. Scientist Lynne Page Snyder explained, "Farmers and residents in Lincoln and Forward Townships, downwind of the Clairton Works, reported experiencing the mill's effluent as a sticky tar that burned skin during harvesting, sickened livestock, ate away at tin roofs, and weakened lungs to pneumonia and influenza."

The conditions soon became unbearable, prompting a number of areas residents to join Gliwa's suit against US Steel. Until the Meuse disaster Pinkasiewicz had been pursuing, for the most part, a traditional legal route of suing for personal damages already suffered. After Meuse, though, and spurred on by the lingering financial effects of the Great Depression, Pinkasiewicz pushed forward on an unusual approach. He fused a personal damage claim with a breach of contract claim.

"This unconventional claim," wrote Snyder in her doctoral dissertation on the Donora tragedy, "framed the argument that United States Steel derived its profits from the use of the shared atmosphere as a dumping ground for wastes, in effect extracting profitability from what was termed the damaging or 'atmospheric depasturing' effect of effluent."

Pinkasiewicz believed that framing the suits that way would allow the farmers to claim part of the income generated by the coke plant. He based the assertion on the legal tenets for a rental agreement. He wanted to show that each plaintiff's "body, respiratory, and blood circulating systems" were being used as a kind of space in which the mill's effluents were "living." As such the mill should be made to pay for the "rental" of that space. The legal briefs of *Gliwa v. United States Steel Corporation* made clear that the farmers had no interest in closing down Clairton Coke Works, but rather wanted only to stop the mill from emitting toxic gases.

———————

US Steel filed a motion to dismiss the initial complaint, and the district court complied with the dismissal, but did so "without prejudice," meaning that the complaint could be amended and resubmitted. Pinkasiewicz did exactly that, amending the first complaint and resubmitting it. US Steel filed a motion to dismiss again, and again the court complied without prejudice. After several rotations on this legal gerbil wheel, the district court dismissed the case *with* prejudice.

At that point the case was appealed to the US Circuit Court. Pinkasiewicz submitted his case, along with sixty pages of supporting "scientific and industrial" data. Judge V. B. Woolley, writing for the appellate court in a decision on April 18, 1932, described numerous deficiencies in the plaintiff's case, often sounding like a disappointed parent scolding a child.

Judge Woolley first admonished the plaintiffs for submitting too many similar causes of action, legal documents that provide a set of facts that would allow a suit to proceed. The case at that point had been amended so many times that Woolley had trouble determining the main cause of action from the six submitted. Woolley chose to respond only to one, saying that the others were essentially the same, just worded differently. "We shall for convenience direct and limit our discussion to the second amended bill, keeping in mind the requirement of good pleading, long in force and emphasized by equity rule 25, that a bill shall contain 'a short and simple statement of the ultimate facts upon which the plaintiff asks relief.'"

It was that phrase in rule 25, "short and simple," that drove the final nail into the case's coffin. The court rejected the second cause of action purely on its length, saying, "This statement of a claimed cause of action, whatever may be its nature, clearly violates equity rule 25, for it is far from 'a short and simple statement of the ultimate facts,' but is on the contrary a lengthy recital of evidence, relevant or otherwise." Woolley then went on to reject the other causes of action because they were based on the second cause. That is, as soon as the second cause was rejected, the others would have to be rejected as well.

Ironically, the judge noted that the clearest indication he could find of a legally sustainable, short and simple cause of action was the one put forth in Pinkasiewicz's initial pleading. Woolley noted that a letter written by the plaintiffs to officials at the mill explained the strategy that Pinkasiewicz had based on the tenets of a rental agreement. The farmers' letter claimed that the farmers were now "leasing" their land to the mill "as a dumping ground for waste fumes." The farmers were thereafter charging the mill an annual rent of one thousand dollars per acre. They declared that "the first next succeeding use by you [the mill] of any of this land will be deemed [by the farmers] a full acceptance of these terms and conditions." If the mill continued to emit waste fumes, that "contract" could then be considered by the farmers to have been breached.

"When the plaintiffs dropped this ground of action from their subsequent bills," Woolley wrote, "they, nevertheless, continued and repeated the same prayers for relief, hence the difficulty in understanding their application to the six causes of action last pleaded."

Pinkasiewicz had altered his pleadings to obtain the best possible outcome for his clients. He had gone above and beyond by providing a robust pool of supporting citations for data on air quality in general. However, he lacked the scientific data needed to support specific damage claims, data not yet available. Results from studies of the Meuse tragedy had yet to be written, and US studies on the effects of air pollution on human health were basically nonexistent. Even the term "smog" wouldn't make it into the national vernacular until the early 1950s.

Without precise information on the health effects of the coke works' effluents, Pinkasiewicz was forced to improvise. He enlisted the help of four members of the farmers' committee: George Gliwa, Charles Faust, former Clairton Works chemist Girard Leech, and an industrial chemist named Edgar Patterson. Together they taught local farmers how to take rudimentary air and soil samples at certain locations on their farms. They then presented their findings to the circuit court. One of the committee members wrote in the brief, "It is not to be contended that all those tests are absolutely accurate; but they are reasonably accurate, and just a[s] reliable as those which chemists give for interested corporations in litigation."

His sentiments might have been on target, but US Steel had no trouble finding its own scientists, each well known and highly credible, to find fault with the committee's "reasonably accurate" results. The company even built an agricultural testing station of its own, adjacent to the Gliwa farm, and operated it twenty-four hours a day with three shifts of scientists and their assistants. Pinkasiewicz's rather homegrown tests were no match for US Steel's expert witnesses and their more exacting test results.

The reality is that neither the district court nor the circuit court could have decided much of anything differently. Judge Woolley even praised Pinkasiewicz, writing at the beginning of the court's decision, "It should be understood that nothing we may say will be in criticism of plaintiffs' counsel, whose great industry and earnest belief in his case doubtless moved the trial court to consider it with much labor and great patience."

The farmers were plainly outspent at every turn and overwhelmed with so many motions from the defendants that even Pinkasiewicz, a persistent and creative lawyer, couldn't slay Pittsburgh's grand Goliath, US Steel. The company had brought to bear on the lone Pinkasiewicz a powerful Pittsburgh law firm: Reed, Smith, Shaw, and McClay. The firm had been founded in 1877 by James Hay Reed and Philander Chase Knox. Reed would go on to become a circuit court judge himself, and Knox would serve as secretary of state for President William H. Taft, an assistant US attorney under two presidents, as well as a US senator and even a candidate for president. Reed and Knox were both members of the infamous South Fork Fishing and Hunting Club and, as the club's attorneys, had successfully defended South Fork against a variety of lawsuits filed in the aftermath of the Johnstown Flood in 1899. Although both men were deceased by the time the Gliwa lawsuits were brought forth, their legacy was nevertheless carried forth in the persons of John J. Heard and John C. Bane Jr., the skilled attorneys who successfully defended US Steel in the Gliwa case.

Regardless of the quality of Pinkasiewicz's pleadings or the accuracy of his test results, the case was undoubtedly doomed to fail without definitive scientific evidence that pollutants floating in the air could cause direct harm to someone's health. That evidence wouldn't exist until the nation began studying in a comprehensive way air pollution and its effect on health, and that wouldn't happen until after the last victim of the deadly Donora fog took his final breath.

———————

US Steel faced many other lawsuits during the 1930s, keeping the Reed, Smith team gainfully employed and richly compensated throughout the decade. The company often lost at trial but won on appeal or, with almost unlimited cash available, kept the cases dragging on for so long that the original plaintiffs—the farmers, mill workers, coal miners, and common laborers who suffered the effects of mill emissions—could not sustain their efforts.

The company didn't win all of its air pollution cases, however. US Steel had lost a case in 1920 that actually set the stage for its win in the Gliwa case and the many others that followed. That case involved a married couple suing their malodorous neighbor, the Donora Zinc Works, for damages to property and health. Frank and Mamie Burkhardt had built a house in 1904 in the north end of town, not far from the river. Frank worked as a coal miner, Mamie as a mother and homemaker. The couple planted trees, vines, and shrubs around their new home. By the time the zinc smelter began operations in 1915, the couple were raising three children, ages eight, ten, and eleven, in a home they could be proud of.

The zinc smelter was built on land almost directly across the street from the Burkhardts', about a par-four golf hole away. Normally westerly winds would carry smoke from the mill's chimneys across the river toward Webster, but when the winds shifted, the smoke passed over Donora.

Mamie had been a healthy young mother when the mill was built, but soon began experiencing frequent headache, sore throat, hoarseness, and "a soreness through her lungs followed by a hacking cough." Her health declined rather quickly, and she lost twenty pounds in a year. Early in the winter of 1916 Mamie visited her sister in New York and stayed six weeks. During that time her symptoms disappeared and she gained back the twenty pounds she had lost. On her return home, however, she became sick again, exposed once more to smoke from the zinc smelter. Frank Burkhardt, having watched his wife suffer and the healthy plants around their home shrivel and die, decided to sue the owners of the mill.

In the meantime Frank and Mamie uprooted their family and moved a few miles north, to Elizabeth, a village a mile or so upriver from, ironically, the Clairton Coke Works. Whether the couple was aware of the sticky-tar air emanating from the chimneys there has not been recorded.

The Burkhardt case finally made it to trial in 1919. The jury decided that damages had in fact occurred and that zinc smelter effluents were most likely to blame. The court awarded the Burkhardts five hundred dollars in damages, an amount equal to about $6,500 today. US Steel appealed, claiming that the case should have been thrown out at the start because Burkhardt failed to provide any scientific data to back up the claim that the zinc smelter was solely responsible for the damages suffered.

The appellate judge, John Head, was not impressed. He wrote that Mamie Burkhardt had described the changes in her health "so plainly that it would be to deny the common experience of men to say that an inference of a causal relation between the two facts was but a mere guess or a vague conjecture." In other words, US Steel's case was pure poppycock.

Judge Head then affirmed the trial court's decision, and US Steel's mighty legal team exited the courtroom beaten. From that point forward US Steel would develop scientific data of its own to defend itself in future court cases, including those that would arise in the months after Donora's deadly smog.

PART III

FOG ROLLS IN

17

THE DAYS
BEFORE

IT WAS DELIGHTFUL IN DONORA TWO WEEKS BEFORE HALLOWEEN IN 1948. Daytime temperatures hit seventy-one degrees on Saturday, October 16, and nearly eighty degrees on Sunday. Nighttime temperatures plummeted each day, typical of fall in the northeastern United States. Folks would take to layering their clothes for this kind of weather, pleasant during the day and downright chilly at night.

Much was happening around the world that month. Germans convicted of war crimes were being executed, the latest of them on October 22, when ten Nazi SS officers were hanged in Lansburg, Germany, for having killed "without mercy" Allied fliers shot down over Germany. On October 16 a "high Israeli source" predicted that a political solution to the future of Palestine would be reached "within two months." Casey Stengel took the helm of those damn Yankees on October 12, declaring later that the secret to being a good manager was "to keep the guys who hate you away from the guys who are undecided." Dominating the headlines throughout October, though, were the names Truman and Dewey.

Incumbent President Harry S. Truman was campaigning against his principal challenger, New York Governor Thomas E. Dewey, in the run-up to the election on November 2. The race wasn't supposed to be a race at all. A Gallup poll that month showed Dewey with a four-point lead over Truman, whom the *Chicago Tribune* had famously called a "nincompoop." The Roper polling company had given up the race as well, going so far as to stop surveying prospective voters in September. None of the polls, however, took into account the effectiveness of Truman's "whistle stop" campaign, and when the election results were announced, Dewey most certainly did not defeat Truman.

But in late October the election was still up for grabs. Truman had spoken to a huge crowd at Chicago Stadium on Monday, October 25, and hammered at the communist threat, saying, "We must not imagine, just because we love freedom, that freedom is safe—that our freedom is safe. Eternal vigilance is still the price of liberty." Dewey spoke in the same arena the next night and, comfortable with his lead, spoke largely in generalities. "To me," he told the sellout crowd, "this is more than just a campaign to win an election. It is a campaign to strengthen and unite our country to meet the challenge of a troubled world."

Neither candidate spoke about air pollution. Whether Dewey had known about the Meuse tragedy of 1930 is unknown, but Truman probably did. Back in the fall of 1918, Truman had just been made a lieutenant in the army when he led his troops on horseback into the Argonne Forest, east of the Meuse River in France, entering into the largest military operation in history to that time, the Meuse-Argonne Offensive. Twenty-six thousand US soldiers died in the battle, the last major battle of World War I.

Twelve years later, in 1930, Truman was seeking and gaining reelection as the Eastern Jackson County judge, holding court in the courthouse now named in his honor, in Independence, Missouri. Truman would have been keeping himself up to date with world news and almost certainly would have spotted any headline with the word *Meuse* in it. When Truman stood before the Chicago crowd in 1948, he had no way of knowing that he would soon face a smog tragedy here at home, and he would be compelled to do something about it.

The weather was cooler than usual when the last full week of the month began, with mild winds and mostly cloudy skies on Sunday and Monday, October 24 and 25. Fog settled in the valley both mornings, as it typically did in October, and dissipated after a few hours. Tuesday saw fog and near-freezing temperatures in the early morning and then steadily warming air into the early afternoon, typical for Donora in the fall. Tuesday's morning fog, though, didn't lift. It stayed all day and all night.

To residents it was just another foggy autumn day. They walked to work, cleaned their homes, made their lunch, shopped in town, and did all the everyday tasks they normally would have. At least one young woman cleaned her house that Monday. She also washed the kitchen furniture and put plastic covers on the living room couch and chairs, trying to keep them somewhat clean. The woman was a homemaker and mother, and loved every minute of it.

Gladys Schempp (née Balmer) was a brown-haired, blue-eyed beauty, with a sweet smile and slight cleft in her delicately shaped chin. She was tiny: five feet, two inches tall and, after graduating from nearby California Teachers College in California, Pennsylvania, in 1939, weighing just ninety-five pounds. She finished fourth in her high school class of 194 and was quiet, serious minded, and kind but stern. She would earn a master's degree from Duquesne University in her fifties and then work as a librarian for the Ringgold School District until her retirement.

Gladys was twenty-two when she married Karl William Schempp Jr., 24, known to everyone as Bill, on September 22, 1941. Gladys had started writing a diary when she was ten and got hooked. She wrote in her diary every day until the day before she died, six days before she would have turned ninety. After she and Bill married, Gladys's diary entries reflected the life of a young woman doing everything she could to create a happy, loving home. Her entries were consistently short and to the point, often in the extreme. Typical entries read along the lines of "Played cards all day and nite—lost 75¢"; "To Mariann's all day to paint guest room / Nite—Peggy Dean's to make favors for Aux banquet"; and "fireman's picnic." Gladys apparently didn't believe in periods; few exist in the thousands of entries in her many diaries. Hyphens, though, abound.

Gladys lived like she wrote. She was a to-the-point woman who strove to make every day count. A period meant an ending, and Gladys would have none of that. Hyphens were just tiny breaks between each life event.

Let's go! they seemed to say. *Keep moving!* She was constantly busy—cleaning the kitchen and then rushing to a meeting downtown, then to lunch with Bill, back to clean the bathroom, then to dinner with friends, and finally back home for an evening of cards.

Gladys managed the house throughout their marriage and quickly discovered that she needed to manage her husband as well. Where Gladys was quiet and introverted, Bill made friends all over town. Gregarious to a fault, Bill was forever being stopped on the street to chat. It was Gladys who had to find him and bring him home so they could go to whatever gathering came next. Gladys paid all the bills, and Bill was just fine with that. "He just couldn't identify the importance of money," recalled the couple's adopted daughter, Annie. Her dad would much rather be working, fighting fires, pulling victims from cars after an accident, liberating frightened cats from trees, and gabbing with his fellow citizens downtown.

That Monday morning, October 25, Bill left for work at the Donora post office around 7:30. Gladys put on an apron and began cleaning. Neither of them had any idea that morning, nor did any of the thousands of Donora and Webster residents, what tragedies lay ahead from a slight change in Tuesday's weather. And they certainly didn't know that Bill would emerge from that historic weekend as one of the town's true heroes.

18

THE FIRST
DAYS

JOURNALIST BILL DAVIDSON WROTE FOR YANK, *THE ARMY WEEKLY*
during World War II and on his return from the war settled in Los Angeles, where he would begin writing for such national magazines as *Collier's,*
McCall's, Good Housekeeping, and *Ladies' Home Journal.* A talented writer,
Davidson was assigned to write a story about the dangers of air pollution in the United States, particularly the effects of sulfur compounds on
human health.

His article for *Collier's,* a weekly magazine, was called "Our Poisoned
Air." It discussed how sulfur had affected people living in the Meuse Valley
when tragedy struck there in 1930. "The coal used around Liège has a high
sulphur content," Davidson wrote. "This sulphur, when burned, produces
a gas called sulphur dioxide, tons of which enter the air in smoke. During
the period of the poison fog, this sulphur dioxide became locked down
in the valley because a meteorological 'roof' of cold air settled on the surrounding hills and prevented it from escaping."

The article then shifted to a discussion of sulfur dioxide pollution in
such cities as New York, Pittsburgh, and Los Angeles, singling out Los

Angeles as being especially vulnerable. Davidson wrote, "Sometimes the 'smog,' as it is called, sends coughing, weeping people streaming homeward from their offices. Los Angeles burns high-sulphur-content California oil in its factories and oil refineries. Every day, some 800 tons of sulphur dioxide pour uncontrolled into the city's atmosphere."

He went on. "The horrifying thing about this is that . . . more than three days of the same set of conditions could cause the same disaster again in any industrial area. The miracle is that it *hasn't* happened in the United States; for it is American habit to poison our air as flagrantly as we have poisoned our waters."

Davison ended his piece with a powerful question: "Are we Americans waiting until we experience a Meuse Valley disaster of our own?" In a breathtaking twist of timing, the article appeared on Saturday, October 23, 1948, just three days before the start of the worst air pollution disaster in American history.

"A fog as thick as Dingbat's skull covered Pittsburgh this morning," read the Donnie Dingbat column in the *Pittsburgh Press* on Tuesday, October 26, 1948. Donnie Dingbat was a cartoon bird that began gracing the paper's front page in 1941, shortly before the nation's entry into World War II. Two years before, with war raging in Europe, President Franklin D. Roosevelt had instituted restrictions on the publication of detailed weather forecasts, fearing that enemies might use the forecasts to plan attacks on US soil. In response, famed newspaperman Edward T. Leech asked the *Press*'s artist, Ralph Reichhold, to "create some kind of animal or bird to brighten a dull weather story." Twenty minutes later Donnie Dingbat was born.

The sarcastic, pun-loving bird gave weather forecasts and current conditions a comic spin, letting readers know what kind of weather was afoot without being too specific. Donnie's forecast for that foggy October day continued: "The relative humidity was 99 percent this morning, about as high as it can go without citizens walking around underwater."

When the relative humidity, or dew point, hits 100 percent and meets cool ground, fog forms. Fog that morning covered an enormous region, extending from Lake Erie and southwestern Ohio, across western Pennsylvania, and down into West Virginia. A large low-pressure weather system had passed through the northeastern United States into the Atlantic in mid-October. High-pressure systems developed in its wake. One high-pressure system that had formed over northern Kentucky and southern Ohio gradually expanded until it covered the southwestern-most tip of Indiana, most of Kentucky, the northwestern end of North Carolina, and all the way up to Buffalo and western New York.

That change in systems led to changes in wind flow. Winds that develop in a low-pressure system (north of the equator) tend to flow toward the center of the system, where pressure is lowest, and blow in a counter-clockwise direction. Winds that develop in a high-pressure system do the opposite. They tend to flow away from the center of the system, where pressure is highest, and blow in a clockwise direction, known as an anticyclonic flow. Anticyclonic flows were responsible for the stagnant air in the Meuse Valley temperature inversion, and were again for Donora's moribund air.

The high-pressure system that covered the entire mid-Atlantic in late October would move lethargically eastward, but fog throughout southwestern Pennsylvania would persist, particularly in Donora. Warming temperatures or even a steady breeze could have dissipated the fog during the day, but neither was on the horizon.

Donorans didn't care a whit what the fog would do. They lived in a valley, and fogs were just a fact of life there. The citizenry went about their day as they would any other. Gladys Schempp's day was filled, and without cleaning duties for a change. That afternoon she baked a cake, then went shopping in West Newton, about five miles east of Donora. Her mom, Anna, visited for dinner. After she left, Gladys finished her day by ironing Bill's work shirts.

Anyone reading that day's *Daily Republican* would have learned that middleweight Lee Sala, a Donora native, had won his match against Brooklyn boxer Tony Demiccio in a match in Toledo, the body of legendary slugger George Herman "Babe" Ruth Jr. had been laid to rest in a hillside grave in a cemetery in Hawthorne, New York, and two "thugs" had stolen $150 from a store owner in Pittsburgh.

Tuesday was just an average day in Donora and vicinity, when nothing particularly special happened. Nothing, that is, except a confluence of weather conditions that would place an environmental lid on the valley. No one knew then how many days the fog would last, nor how long the lid would hover over the valley. Yes, Tuesday was, almost, an average day in Donora.

Donorans woke up Wednesday morning to fog clinging to the valley like shower steam in a fanless bathroom. If they had thought about the fog for even a moment, if they had regarded it a bit more closely, they might have noticed that it was thicker than yesterday's fog, and darker, certainly darker.

Donora's mayor, August Zephirin Chambon, probably greeted people he met on his way to work that morning with a comment on the lingering fog. Chambon owned a moving and storage company in Monongahela and had been serving as Donora's mayor since 1941. Born in Bessèges, a tiny community in southern France, Chambon had dark hair that he combed straight back, with no part, and a wide forehead. He had a stern look and double chin and could have passed as a Jimmy Hoffa lookalike. Chambon had been a fixture at the microphone for town events for many years, and pretty much everyone knew him.

Gladys and Bill Schempp picked walnuts in the afternoon and took a friend to Capone's, a local Italian eatery, for dinner. They sat in one of the booths, as they typically did. Bill ordered spaghetti. He always ordered spaghetti. Gladys was more adventurous, changing her meal order each time. She never changed her order for dessert, though. She always ordered a slice of Capone's delicious cheesecake. "She loved the cheesecake there," said Annie Schempp. "Cheesecake was a rarity in restaurants back then."

At some point during the day Bill had spilled a quart of paint at home, which could not have pleased his loving wife. She recorded the incident, though, as dispassionately as she did any other: "Bill spilled 1 qt paint." Gladys recorded the facts and nothing but the facts.

Newspaper headlines Wednesday afternoon noted Dewey's lead over Truman in eighteen out of twenty-four northeastern states. The *Daily Republican* announced details of the Monongahela Halloween parade, which would start at 7:15 p.m. Thursday, the night before Donora's parade. The Monongahela High School football team, the Wildcats, were practicing "serious drills" for its "toughest assignment so far this season," a Saturday game against the school's rival, the Donora Dragons. The Dragons, whose colors were orange and black, had been named for the way Donora's blast furnaces and open hearth furnaces appeared at night.

The Monongahela coach, Ben Haldy, admitted that beating the Dragons would be "awfully tough since they're so big." The Wildcats would face a Donora team that, per lineman, outweighed them by an average of twenty-six pounds. "Such a weight advantage could easily be the deciding factor in the game," said the *Daily Republican*. As it turned out, weight wasn't the deciding factor at all, not even close.

The fog that formed on Tuesday settled into the valley like a lazy cousin on a three-week visit. The mill's many chimneys continued to pour forth toxins in its smoke all night, just as they had been doing every day and night for decades, fog or no. Rather than being caught on the wind and transported throughout the region, smoke from the metal factories—and the toxins contained in it—now were trapped. The valley's walls hemmed in the fog and smoke from either side of the river. The temperature inversion blocked the release of fog and smoke upward, and the horseshoe bend in the Monongahela stymied their release to the north or south. The valley had become by Thursday a lidded mixing bowl, continually blending discharge from cars, trucks, trains, and mill chimneys into what was rapidly becoming a sickening brew of dark gray muck.

Even the dark gray color of the air was different. Normally the color of the smoke varied throughout town. Near the blast furnaces at the south end, smoke tended to look black, largely due to coal being used as fuel. In the middle of town, where the open hearth furnaces were, smoke tended to be reddish in color, from the iron ore being broken down. Finally, at the north end of town near the Zinc Works, smoke tended to have a yellowish tinge, the result of sulfur-containing fumes being given off during smelting. With a lid now over the valley the smoke began to blend into a deep, gray mélange of poisonous smoke.

Had factory smokestacks been taller than 250 feet, the approximate height of the lowest level of the inversion layer, the effluents coming from them might have spewed into the atmosphere, where they would have become diluted and dissipated over the region. The tallest smokestacks in Donora, though, were just 150 feet high. Without wind, even a slight breeze, there could be no upward movement of smoke from the plants. So smoke, soot, dust, and toxic gases from the plants, trains, and vehicular traffic in the valley continued to mix into the fog.

It seems utterly unlikely that mill owners didn't know how important stack height was to the surrounding communities and, more important, their bottom line. A smelter operator in Montana, the Anaconda Company, constructed a three-hundred-foot stack at its Washoe copper smelter in 1902. Farmers and ranchers near the smelter found that they were losing crops and livestock due, they claimed, to "smelter fumes and poisonous ingredients" contained therein. Fred J. Bliss, on behalf of area residents, sued Anaconda for damages.

Although Bliss, not surprisingly, lost his claim, Anaconda leaders took the lesson to heart, and in 1917 the company removed the smaller stack and built in its place a stack that reached 585 feet into the air, a chimney famously known as the Anaconda Stack. Numerous other smelters at the time also constructed smokestacks more than three hundred feet high, but owners of the zinc smelter in Donora chose otherwise. Andrew Mellon's key lieutenants must have known about the trend toward higher stacks; almost certainly Donora's mill officials did. Why they chose to ignore that trend is unknown, but the decision cost Donora dearly.

Bernardo Di Sanza, track foreman for Donora Southern Railroad, began feeling unwell around nine o'clock Thursday night at his home. He and his wife, Liberata, lived at 337 Third Street, across from a small wooded area known now as Cascade Park. The Di Sanza house was long and narrow, with two bedrooms and one bathroom. Di Sanza was a healthy, happy, some would even say jolly man and could be found nearly always smiling and laughing, his now-gray mustache bobbing up and down.

Tonight was different. Tonight he was not smiling or laughing. He was coughing a great deal and had a hard time breathing. He was never sick, not ever, so this was something that neither he nor his wife were accustomed to. He fell asleep that night hoping he would feel better in the morning.

The congealing fog in nearby Donora didn't stop fans of the Monongahela Wildcats from buying tickets to the game on Saturday. There was already a line at the Pulaski News Store in downtown Monongahela when

745 tickets went on sale at two o'clock. By four o'clock every ticket had been sold. It seemed that nobody wanted to miss the big game.

Donora was rife with football fever as well. Donorans loved their sports, especially football and baseball. Stan "The Man" Musial by 1948 had become a huge star in baseball and had just finished a tremendous season, hitting a career-high thirty-nine home runs. Musial finished first in voting that year for the National League's Most Valuable Player award, well ahead of standout pitcher Johnny Sain, famed shortstop Harold Henry "Pee Wee" Reese, and slugger Ralph Kiner, who hit "only" forty home runs in 1948. (He had hit fifty-one the year before and would hit fifty-four the year after, becoming the first National Leaguer to hit more than fifty home runs in a season twice.) Donora would later produce two more baseball legends, Ken Griffey Sr., whose father worked for a time at American Steel & Wire, and Senior's son, Ken Griffey Jr. The two Griffeys would become the first father-son duo to play professional baseball on the same team at the same time.

Donora was equally proud of its football heritage. The Dragons were then known as a challenging team to beat, year in and year out. The team had played in the Western Pennsylvania Interscholastic Athletic League, composed of about 160 schools. Donora had won back-to-back championships in 1944 and 1945, outscoring their opponents by a stunning 282 points in 1944 and by 284 in 1945. Both teams featured all-state fullback Daniel "Deacon Dan" Towler, arguably the greatest football player ever to come out of Donora. Towler was picked up by the Los Angeles Rams and, together with such stars as Norm Van Brocklin and Elroy "Crazy Legs" Hirsch, played an integral role in the Rams' powerful running and passing offense. Towler would score the game-winning touchdown in 1951's World Championship game, the precursor to today's Super Bowl, and would be selected for the All-Star team three years in a row.

Other Donora football virtuosos included quarterback Arnold "Pope" Galiffa, whose photo graced the May 2, 1949, cover of *Life* magazine; Galiffa's favorite running back, Roscoe Ross; and Lou "Bimbo" Cecconi, another quarterback who would play both football and basketball at the University of Pittsburgh. So many outstanding athletes came from Donora that the town would, in the late 1940s, begin calling itself "Home of Champions."

Donorans on the Thursday before Halloween must have felt supremely confident that the Dragons would win such an important game, and hundreds were planning to attend. If any fans developed a cough in the days leading up to the game, they probably would have thought they were coming down with a cold. Or perhaps they would have thought they were allergic to some autumn pollen. If, however, they had blamed the fog for their woes, they would have been correct. Toxins continued to spill into the air and become more concentrated with each passing hour. People were beginning to feel the effects.

19

FRIDAY

PAUL GARRETT HAYES WAS A YOUNG FAMILY MAN. ROUND, BESPEC-
tacled, and friendly, he taught physics at Donora High School. Hayes
kissed his wife, Veronica, goodbye and set off for work at 7:30 Friday
morning. Within minutes he became short of breath and nearly choked.

Hayes suffered from asthma and had been having frequent attacks
lately. His doctors had told him a few months before that smoke from
the mills was causing the attacks and that if he wanted to survive past his
twenties, he should move out of the valley. Being a man of science, and
perhaps a bit stubborn as well, Hayes decided instead to install a device, at
great expense, that would clean air inside the house. The device, an elec-
trostatic air filter, applied a static charge to air being drawn through the
filter. The charge caused particles in the air to stick to the filter rather than
flowing around the house.

Outside in the thick fog, however, there was no filter. Hayes felt
another asthma attack coming on. He turned on his heels and headed
back home. Once inside, his breathing eased. He would not teach school
that day.

Across the river a small group of miners were descending into a coal pit. They had been having trouble breathing since the day before, but now, down in the pit, their breathing slowed and they felt better, stronger.

Housewives Betty Crafton and May Ridgely were sending their respective children to school at around the same time. As they did they found black, sooty mucus running from each child's nose. Sooty mucus was nothing new to children in Donora. Betty's and May's kids often came home from school with black nostrils and black at the corners of their mouth from all the soot and dust in the air. But to have that in the morning? Something seemed not quite right.

Dr. Ralph Koehler gazed out his bathroom window at 8:30 a.m. and watched as smoke from a locomotive near the river rolled to the ground. The smoke looked to him like oil flowing down the engine, and rather than blowing away, the smoke just lay there, unmoving. The scene mystified Koehler, who had a taut personality anyway. He believed strongly that Pittsburgh needed better smoke control, but at least it had *some*. Donora doesn't have *any* smoke control.

Born August 19, 1900, in tiny Reynoldsville, Pennsylvania, seventy miles northeast of Pittsburgh, Ralph Waldemar Koehler was one of Donora's eight family doctors. He was the second son of Roman and Jennie Koehler. Roman served as managing editor of the *Donora American* newspaper and was Donora's first justice of the peace. Son Ralph graduated from the University of Pittsburgh School of Medicine and set up practice in Donora.

That morning, while Koehler bemoaned the lack of smoke control to himself, he became more and more upset. He had high blood pressure and an ailing heart, so he knew not to exert himself too much. He finally sat on the edge of the tub to rest. After a few moments he stood and walked to his room to dress and start another day. By that time many Donorans had been up for hours. They had awakened with a cough and shortness of breath. Some of them would soon head to Koehler's office at Fifth and McKean.

The office manager, Helen Stack, was at the office and cleaning an unusually scummy dust that seemed to cover everything. She wiped down the desk and chairs in Koehler's office, and then did the same in

Dr. Edward Roth's office. She washed the examination tables, sinks, and waiting room chairs, and then, a little after nine, sat in her chair to light a cigarette. The cigarette tasted terrible, awful. There was a sweetness to it, but also something putrid. *It tastes like the fog,* she thought, *only ten times worse.*

Stack took another puff, perhaps thinking it might help ease her cough, but instead she erupted in a coughing spasm. She quickly snuffed out the cigarette, crossed to an office sink, and drank a glass of water. She had smoked a cigarette after breakfast at home, and it tasted fine, but this one was different. *Maybe it's the fog,* she thought. *Maybe it's worse down here on McKean than at home.* Then again, she probably just had a cold.

She could hear talking and clanging outside and figured that workmen on McKean were hanging decorations for the Halloween parade that night. The chamber of commerce put on the parade every year, and if the fog was dangerous, surely the chamber would cancel it. That decorations were being hung told Stack that the parade would go ahead as scheduled, so how bad could the fog be?

Milton Mercer Neale, known as M.M., had arrived in Donora on September 30, 1909, when he was twenty-three. He graduated at age nineteen from the Virginia Polytechnic Institute in Blacksburg with a degree in civil engineering and proved so bright and capable to American Steel & Wire officials that he was named superintendent of the mill ten years later. Neale made a name for himself over the years and by 1948 had become well known throughout the valley.

Neale, with glasses, a balding pate, and a sober countenance, fit the look of a midlevel corporate executive to a T. Superintendent Neale was responsible for all operations of the zinc smelter, overseeing everything that went on there, inside and out. He was energetic and ambitious. He was widely respected and well appreciated for the plentiful worker benefits he initiated. For instance, Neale established a credit union at the mill, an athletic association, and a mutual society. He organized a band, an orchestra, and a men's chorus. He served as chairman of Donora's World War II Chest Fund and treasurer of the World War II Honor Roll. He became a charter member of the Monongahela Valley Country Club and an honorary president of the Spanish Benefit Society, which feted him in

January 1950 for "his efforts in behalf of community betterment." Neale also played leading roles in the local and county Boy Scouts Association. He seemed, by all accounts, to be a genuinely nice guy.

Neale lived at the corner of Tenth and Thompson with his wife, Alma, and as residents they had been experiencing the thickening fog themselves. Neale decided to conduct a routine inspection of how well smoke from the zinc smelter was being dispersed from its stacks. It was noon when he noted that the stacks were not emitting an excessive amount of smoke, he believed, nor did the emissions look in any way unusual. He found no reason to dead-fire the furnaces. The term "dead-fire" means to halt the introduction of raw materials, the charge, into the furnace. Without a charge to keep the furnace going, the temperature inside the furnace would slowly drop. The lower temperature would then be maintained to prevent the brickwork inside the furnace from cracking due to thermal shock.

After Neale examined smoke from the stacks, he ate lunch with two of his managers. One of them reminded Neale of the Meuse disaster in Belgium and how many people had died in that fog. Neale explained that the fog was darker downtown than it was around the zinc smelter—a possible, but unlikely, claim—but that he would check the stacks again that evening.

Koehler and Roth arrived around noon from their morning rounds at Charleroi-Monessen Hospital, a sturdy, four-floor brick building with eighty-five beds around the hill in North Charleroi. Patients started streaming through the doors shortly before one o'clock, when office hours officially started. Stack moved the patients in and out as quickly as the doctors could see them, and for the next two hours the office ran normally.

Koehler generally left at three o'clock for a respite at home. He was just about to leave when Stack announced that one of their regular patients with asthma had arrived, gasping for air. Koehler listened to the man's chest and heard wheezes typical of an asthma attack. A wheeze is a high-pitched whistling sound caused by a narrowing and spasming of the airways, a condition known as bronchospasm. In bronchospasm the amount of air that can pass into and out of the lungs is greatly reduced and, if left untreated, can stop the heart and lungs from working.

Koehler also noted that tissues in the man's neck and shoulders seemed to sink into his chest with each inhalation, a typical occurrence, called retraction, in a patient in the throes of an asthma attack. He stepped to a cabinet and prepared a syringe of Adrenalin to open the patient's airways and help him breathe better. Then he injected the medicine into the fleshy part of the man's upper arm. Within minutes the man's breathing eased, the retractions disappeared, and the wheezing quieted. After the patient left, Koehler headed home, not feeling well himself.

An elderly steelworker wobbled through the office door about a half hour after Koehler left. Stack escorted him into Roth's office. Roth gave him an injection of Adrenalin and sent him home. Stack watched the man, feeling only slightly better, walk slowly out of the office into the afternoon air. *Poor fellow*, she thought, *there's nothing sadder than an asthmatic when the fog is bad.*

Moments later Stack heard loud noises on the stairs outside the office and assumed that the poor fellow had fallen down the stairs. Instead she saw a different man slumped over the railing, choking and moaning. He screamed, "Help! Help me! I'm dying!"

Dr. DeWees Englert Brown, who operated a surgical practice across the hall from Koehler and Roth, heard the commotion and ran into the hall. Brown rushed down the stairs to the help the man into his office, then gave the man an injection to ease his breathing.

As Brown was helping the man Stack heard her office phone ring. On the way to answer it she nearly slammed into Roth, who had heard the rumpus and came out to investigate. Stack picked up the phone and said, "Doctors' office, Miss Stack speaking." The man on the other end was screaming, telling her that he can't breathe, that he needed a doctor, that he felt like he was dying. She told the man that Dr. Roth would come right over and then hung up.

The phone rang again. Someone else couldn't breathe. Stack wrote the person's name and address on a slip of paper and hung up. She also noted the time: 3:30 p.m. The phone rang almost immediately, another patient with breathing problems. The rest of the afternoon went exactly that way, patient after patient calling the office, needing a doctor right away. They all had similar complaints: Can't breathe. Choking. Pain in the abdomen. Splitting headache. Nausea and vomiting. Even coughing up blood.

The young assistant was working on her list when Roth walked to her desk, his black leather medical satchel in hand. Roth was a thickset man with black hair and dark, kind eyes behind a pair of spectacles. He had a roundish, almost cherubic face, and as he stood in his hat and coat before Stack, he chomped on a cigar. When Stack handed Roth the list, he scanned it, shook his head in disbelief, and walked out the door.

As soon as Roth left, Stack dialed Koehler's home number, but there was no answer. She didn't have time to try again before the phone rang. Another patient was struggling to breathe. She added the name to a new list, and another call came in. Calls were coming in almost constantly. Every now and then she found a moment to try Koehler's number again, but no one ever answered.

Around the same time that Stack was trying to reach Koehler, the family of Bernardo Di Sanza was desperately trying to reach their family physician, William Rongaus, whose office was three blocks from the Di Sanza home. When they finally reached the office, the receptionist took a message and told them that Dr. Rongaus would get there as soon as he could.

The tiny hospital at American Steel & Wire wasn't as busy as the physicians in town; they recorded just one worker with shortness of breath and a history of asthma. A doctor there, possibly Martin James Hannigan, a full-time physician with American Steel & Wire, treated him and sent him home at 3:45 p.m. Hannigan would be considerably busier tomorrow.

John Robert West was a strong, healthy fifty-five-year-old miner. West lived in Forward, a largely wooded area north of Webster, with his wife, Carrie. They were both born somewhere in Georgia and had been living in the valley for about thirty years. West on this day was walking home from the mine, around five o'clock in the afternoon. The fog wasn't horrible in Forward that day, so he felt reasonably well. Suddenly he saw a "big, black cloud" wheeling toward him.

There had been some valley fogs that bothered him, but never for long. This one, though, looked so ominous that it terrified him. He took off running, but he couldn't outpace it. The cloud swallowed him like an ocean wave, taking the breath from his lungs and burning his eyes. He just kept running, frantic to reach home. He felt sick to his stomach, and his head throbbed with pain. When West at last reached home he staggered through the door and collapsed. His startled wife rushed to his side, surely wondering what on earth had happened. "The black cloud came after me," panted West. "It's going to kill me."

Around the time John West was collapsing, Georgette Chambon was returning home. She had been shopping downtown, struggling to find her way from store to store. Every store she entered was filled with smoke. She finally gave up and headed home. Her mother-in-law, Fannie, wife of Donora's mayor, was home alone today, and Georgette needed to check on her. When Georgette walked through the door she found the seventy-eight-year-old prostrate on the hallway floor. Her face was blue, and she was fighting to breathe.

Mother Chambon's coat was on, and cookies were spilled all around her. Terrified, Georgette helped Fannie to her feet and almost carried her to bed. "I went to the bakery for some cookies," Fannie gasped. "The fog. I couldn't breathe."

Fannie Chambon had experienced lung problems before but never like this. Georgette settled Fannie into bed and then rushed to the phone to call the family doctor, Ralph Koehler. The phone was busy every time she called, but at last she got through. Helen Stack answered and told Georgette that both Koehler and Roth were out. She would add Mrs. Chambon's name to her list, but she had no idea when either doctor would return to the office or call in for another list of sick patients.

Georgette left a message and then, thinking that Fannie needed help straightaway, called another doctor, but he too was out. Finally she reached Dr. Herbert J. Levin, whose office was one block north of Koehler's. Levin told her that he would get there right away, but the trip took him longer than expected. He too was receiving multiple calls and had to decide which patients to visit first.

As soon as Levin arrived and saw Fannie Chambon he knew exactly what was wrong. She was wheezing heavily from bronchospasm and her color was poor. Levin had already treated several patients with bronchospasm, so he wasn't surprised that the elderly Chambon was experiencing it as well. What concerned him was her skin color. Blueish skin indicated that her oxygen level was extremely low. He injected her with Adrenalin to open her airways. She also needed oxygen, but he didn't have any with him.

As Fannie's breathing eased Levin gave her a sedative for sleep, probably either pentobarbital or chloral hydrate, sedatives in common use at the time. Fannie Chambon ended up sleeping sixteen hours straight. As Levin readied to leave, Georgette asked, "Doctor, what's going on? There seems to be an awful lot of sickness going on all of a sudden. I'm coughing a little myself. What is happening?"

"I don't know," Levin said. "Something's coming off, but I don't know what."

20

HEROES AND
VILLAINS

"DOC BILL" RONGAUS WAS FINALLY ABLE TO REACH THE DI SANZA HOME at six o'clock Friday evening. By that time Bernardo was in bad shape.

William Joseph Rongaus had been born to Simplicio and Maria Roncace on April 29, 1914. His parents lived in Amatrice, Italy, a village known worldwide for its eponymous pasta dish, a decadent blend of pork, pecorino, and tomatoes. The family decided to move to the United States and purchased tickets for the steamship *Oceania*, leaving from Naples bound for New York Harbor. They traveled from Amatrice to Naples with their three children: Frank, not quite eight years old; Philomena (everyone called her Helen), five years old; and little Emma, seven months. When they reached Porto di Napoli, however, officials discovered that Frank had a case of conjunctivitis, a highly contagious eye infection, and they turned him away. Simplicio and Maria then had to decide whether to forgo what might have been their only opportunity to emigrate or to leave their oldest child behind, to stay with a relative for who knew how long. Such a dreadful decision for any parent to make.

The couple chose to leave Frank behind and continue their journey.

Family legend has it that Maria had sewn twenty-three dollars into the hem of her skirt as a last bit of savings. The now-smaller family, none of whom could speak English, arrived in New York December 17, 1909. Ellis Island officials anglicized their name from Roncace to Rongaus. The Rongauses eventually settled in Pittsburgh. It would be five years before they were finally reunited with Frank, who by then had transformed from a lad who had only recently learned to tie his shoes into a considerably taller, brand-new teenager.

Simplicio and Maria would have seven more children: Walter, Alley (who died at four years from a severe bout of scarlet fever), William, Jean, Leon (who would go on to operate Donora's prestigious, nationally renowned Redwood Restaurant, a favorite spot of Pittsburgh Steelers players), and Yolanda. William suffered repeated ear infections as a child, which left him with a pesky hearing loss.

William's older brother, Walter, wanted to become a doctor, but the family lacked the finances to pay for his education. So after graduating from Donora High School William took a job at the mills. His income helped the family send Walter to medical school. With Walter on his way, William began his own foray into medicine. He completed his undergraduate work at Washington and Jefferson College in 1940 and was in medical training at Jefferson Medical School in Philadelphia when World War II called.

He finished in just three years, the US Army pushing for shortened educations for physicians. With a college degree under his belt Rongaus was commissioned as a second lieutenant. His registration card noted a 140-pound male, five feet, six inches in height, with blue eyes, brown hair, and a light complexion. It also noted a scar on the "left side of jaw drawn to one side."

Rongaus never saw combat; his hearing loss precluded that. Instead he was assigned an internship at West Penn Hospital in Pittsburgh for the remainder of his stint. The internship changed his life, for it was at West Penn that he met a blond, curly-haired, blue-eyed nurse with a sweet smile named Laura Virginia "Ginny" Edwards. Ginny had attended college to study geography for a teaching career but soon decided that teaching wasn't for her. She switched to nursing, and by the time she met Bill Rongaus she was working as an operating room nurse at West Penn.

The two were smitten. Rongaus proposed to Edwards with a sparkling

FIG. 20.1. William Rongaus on graduation from Jefferson Medical College in 1944. Courtesy of Nancy Rongaus Cherney.

diamond ring. Diamonds were difficult to find at the time because they were critical to the war effort. Producing weapons of war, such as tanks, artillery, aircraft engines, torpedoes, and advanced radar, depended on quality industrial diamonds, so when Edwards's one-carat diamond—with side diamonds too!—flashed in the light, people noticed. Nurses from around the hospital flocked to see the ring and congratulate the newly betrothed.

After Rongaus was discharged he moved to Donora and purchased an impressive home at Sixth and Thompson, directly across the street from Saint Dominic's Catholic Church. The house had been a physician's home and office, so the setup was perfect for Rongaus. He lived in an apartment on the second floor and ran his medical practice downstairs. Ginny and Rongaus finally married in 1954. By that time Walter Rongaus had returned from his stint in the war and had joined William in a joint practice.

Doc Bill, as he was called, was a bit of a character. When he was in medical school in Philadelphia he is said to have hired an organ grinder to play all night outside a competing fraternity during finals week, an annoying, though rather clever, way of interrupting studies of the students inside. Always with an unlit cigar hanging from his lips, Rongaus wore three watches at all times, two on one wrist and one on the other. The sight of them sometimes proved too irresistible for kids in town to ignore. They would run up to him and ask, "Hey, Doc, what time is it?" Rongaus would smile, fling his arms into the air to expose his wrists under his shirt, bring them back down, and respond with an exact average of the time noted on each watch. "He didn't care about time," said his daughter, Nancy. "He just loved watches."

Rongaus might well have been gnawing on a cigar when he arrived at the Di Sanza home. He stepped to Bernardo's bedside and found his patient extremely short of breath. Di Sanza's pulse was racing, the tissues in his neck were retracting, and he had a look of terror in his eyes. Rongaus gave him a shot of Adrenalin, told the family to do their best to let him rest, and hustled away to his next house call. As he trudged up and down the Donora hillside he sometimes caught glimpses of people through their windows, and he began to see how many people were sick. "You could see them," he said, "sitting over their chairs, gasping for breath." Everyone, it seemed, was coughing, short of breath, nauseous, and anxious.

Edward Roth had been making house calls all afternoon as well, cigar in hand. Roth smoked cigars far more often than anyone in his vicinity might have liked. He was a Pittsburgh-area native, born in Homestead to Austrian parents, David and Julia Roth, on April 20, 1907. During the 1930s, when rabid antisemitism reared its vile head throughout the United States, Roth found himself applying to medical school after medical school, each refusing him entry, until finally Washington University in Saint Louis opened its doors to him. He and his family knew that all of those rejections had resulted purely from anti-Jewish sentiment; they were convinced of it.

They were almost certainly right. During the buildup to World War II, many European Jews headed to America to escape persecution and death. The United States rebuffed many of those immigrants, refusing them asylum and turning them away from US shores. In perhaps the most egregious instance, the government, led by Franklin D. Roosevelt, refused in June 1939 to allow 937 passengers on a German ocean liner to disembark in Miami after they had been refused entry to Cuba only days before. Nearly all passengers on the MS *St. Louis* were Jewish. The ship was forced to return to Europe, where 254 of the passengers later perished in the Holocaust. So strong was antisemitic sentiment at the time that Roosevelt, who had been searching for ways to save European Jews, encountered a political system that found it more tolerable to deny asylum to desperate Jews than to admit them entry here.

In any case, Roth graduated from the Washington University School of Medicine and soon found himself hanging a shingle in Belle Vernon first and then in Donora, for the "Practice of Medicine and Surgery," as the diploma in his office read. He met his future wife in Pittsburgh on December 7, 1941, the day Japanese warplanes attacked Pearl Harbor. He might well have been smoking a cigar when the two met. He had, after all, smoked them for years.

Sarah Eva Weiner, whom everyone knew as Sally, did not like cigar smoke. Not at all. Born September 4, 1914, in Pittsburgh, Sally had always disliked men who smoked cigars, and she hadn't particularly cared for overweight men either. Eddie was both. Over the years, when people asked Sally how she and Eddie had met, she would tell them, "I would never marry a fat man who smoked cigars." Then, with impeccable timing, she would add, "So we were married on March 29, 1942."

FIG. 20.2. Edward Roth in his US Army uniform, 1944. Roth served from 1942 to 1946. Courtesy of Jerome Harris.

The couple never had children; Sally's body refused to let her carry a child to term. She was devoted to Eddie, and he to her. She was the tolerant wife of an old-time family doctor, the kind of doctor who made house calls whenever needed, regardless of how many hours he had worked that day or had slept that night, who delivered babies in the dead of night from mothers stretched out on a kitchen table, who stitched the cut knees of kids running around on rocky playgrounds and the eyebrows of rugged mill workers who had gotten into a scrape at a downtown bar the night before. Sally was the proud wife of a man whose medical decisions saved untold lives, whose quiet, assured demeanor put patients immediately at ease, and whose devotion to the community was unquestioned.

Sally might have wished that her husband wasn't such a softie, that he wasn't so generous with his time and efforts, but that just wasn't him. If one of Roth's pregnant patients was ready to deliver but couldn't possibly pay for the delivery, what could he do? "Can't shove it back in," he would say. Many of Roth's poorer patients paid for their office visits with produce from their garden, baked goods, or lovingly homemade knickknacks. Roth accepted the "payments" with gratitude. The man couldn't refuse anyone. So when people started calling his office during the fog, people who weren't even his patients, he responded. Like Doc Bill, Doc Roth set about walking up and down the hills of Donora, visiting one house after another, doing whatever he could to help the sickest residents.

Both physicians treated some "pitiful cases," as Roth called them. They weren't all asthmatics or older patients with lung disease either. Some of them were healthy and had never been bothered by fog before. "I was worried," Roth recalled, "but I wasn't bewildered. It was no mystery. It was obvious. All the symptoms pointed to it, that the fog and smoke were to blame. I didn't think any further than that."

He finished the last call on his list and drove back to the office to replenish the supplies in his medical bag. The fog was so thick on McKean, though, that he almost walked past his own office building. He pulled over and parked the car. When he opened the door he smelled coal smoke and tasted soot in the air. Bitter, dry, disgusting. His chest felt tight. His pulse quickened, a reaction by the body of not having enough oxygen to feed its organs. Roth wondered whether he, too, would soon be wheezing.

As he started climbing the stairs to his office he began coughing. Once inside he coughed and hacked until he started to choke. Then he felt nauseous. Helen Stack wasn't there. *It's just as well*, Roth thought, *I would hate for her see me like this.*

Roth hurried to the bathroom, closed the door, and vomited into the toilet. He flushed, washed his face and hands, and used the wall and chairs to support himself as he walked back to his private office. His hands shaking, he carefully drew up a dose of Adrenalin into a syringe and injected himself with it. Then he collapsed into his chair to rest. Gradually he begin to feel better, to breathe better. He wasn't coughing as much.

Out of pure habit he picked up a cigar and lighted it. He had smoked a few cigars when he was out visiting patients, and they had each tasted terrible, but they didn't make him cough. One puff on this one, though, sent him into a paroxysm of hacking. He crushed the cigar foot in the ashtray on his desk. Then he heard the phone at Miss Stack's desk ring. He was so exhausted from coughing and trudging around town making house calls that he lacked the strength to answer it. He slumped back in his chair. The phone kept ringing with calls, one after the other. Roth ignored them all and closed his eyes to rest.

Down at the zinc smelter, at around six o'clock, Superintendent Neale was walking around his plant, looking. He noted the darkness of the fog and checked how well and in which direction smoke was pouring out of the smokestacks. Again he found no reason to bank the furnaces; the smoke was flowing normally and slightly northward, as before. *If there is a problem with the fog*, he thought, *it's not from us.*

However, a number of workers had been to the mill hospital that month with fog-related complaints, including one who was treated as recently as October 21. The patient that day suffered from asthma and was having difficulty breathing. The physician's note read, "Asthma—breathing in smog."

That entry, with the word "smog" in it, was the first reference to what the "fog" in Donora actually was. A combination of smoke and fog, the word "smog" had been coined by Henry Antoine Des Voeux in a paper submitted to the Public Health Congress in London in 1909. Des Voeux was widely commended for "coining a new term for London fog." In Donora of 1948, though, nearly all physicians familiar with the term were loathe to use it; doing so might imply that the mills were to blame for people becoming sick. One just didn't blame the mills for anything. One only praised them.

Neale had been notified of at least one, and possibly more, zinc plant worker who had been suffering from exposure to smoke from his mill. A year before, on October 20 or 21, 1947, a Mr. Celapino had called Neale to tell him that he and his family had been forced to leave Donora several times that fall due to smoke from the smelter. On a particularly smoky day in February 1941 four boats were damaged in two separate fog-related collisions on the Monongahela. One collision caused the steamer *Clairton* to overturn and one of its barges to sink. Of the river mess the *Pittsburgh Post-Gazette* wrote, "The scene of the accidents is known to river men as one of the worst spots on the rivers. It is a sharp curve over which pours heavy clouds of smoke from the Donora mills, making visibility poor."

Most recently, at least three people had been treated at the mill hospital on October 26, the first day of the smog. Notations on those individuals included, "Headache—Fog," "Dyspnea—fog—taken home," and "Dyspnea due to fog." ("Dyspnea," pronounced *disp-nee-uh*, is the medical term for difficulty breathing.) Neale might not have been aware of those three patients specifically, but it seems likely that he had been notified at some point of the sudden increase in the number of workers seeking healthcare at the mill hospital. It seems disingenuous, then, for Neale to have claimed that his mill couldn't be causing a "problem with the fog."

Neale might have been the nicest guy in the world, but he was a company man. He hadn't reached his powerful position by opposing corporate decisions. He never would have accepted blame on behalf of US Steel, not for the Celapinos leaving town, nor the multivessel collisions. No, the company wouldn't be to blame even if the smog had made him sick too.

21

HALLOWEEN
PARADE

THE PHONE WAS RINGING WHEN HELEN STACK ENTERED. SHE HAD BEEN eating dinner at a nearby restaurant. She took off her coat and draped it over the back of her chair. Then she answered the phone, added another patient's name to the latest list, and hung up. Suddenly she heard groaning coming from the direction of Roth's office. She found Roth melted into his chair, groaning and coughing, his face brick-red and covered with perspiration.

"Oh, my goodness, Dr. Roth, are you okay?" she asked. "What can I do?"

"Nothing, Helen, I'm all right now. I'll get going again in a minute. You go ahead and answer the phone."

Stack was panicked. Her doctor looked like the patients who had been arriving at the office all afternoon, and it frightened her. She finally got through to Koehler, who had been out making his own house calls. Koehler told the young woman, "I'm coughing and sick myself, but I'll go out again as soon as I can."

The phone rang almost as soon as Stack hung up. She answered it, wrote something on a piece of paper, and hung up. When she looked up a patient was standing at the door. He, too, was short of breath and wanted to see the doctor. "The next thing I knew," said Stack, "the office was full of patients, all of them coughing and groaning. I was about ready to break down and cry."

Stack was a receptionist, not a nurse, though she did help the doctors with patients now and then. She didn't have training for this sort of crisis, though, and felt hopelessly overwhelmed. Roth was able to examine and treat a few of the sickest patients, but he felt horrible himself. After scanning the patients remaining in the waiting room, Roth decided that none of them were ill enough to require immediate care, so he decided to make more emergency house calls. He grabbed his medical bag, restocked it, and headed weakly out the door.

All Stack could do now was explain to the patients that nothing more could be done and that they should go home and await a doctor's visit. "I've never felt so heartless," recalled Stack. "Some of them were so sick and miserable."

Around seven o'clock, just as Helen Stack was ushering forlorn patients out of the waiting room and locking the office door, the annual Halloween parade began. Fog or no fog, the parade must go on. People lining the route cheered and yelled as the bands marched by. The fog was so thick, so dark, that people on one side of McKean couldn't see people on the other side. In fact they could barely make out the people marching down the middle of the street. Grace Mills, though, spotted her daughter right away.

Mills and her husband, Alden, lived in Belle Vernon, the next town south of Monessen. They had dressed their five-year-old daughter, Lana, in a costume for the Halloween parade, of which there were four in the area held each year: Donora, Monessen, Charleroi, and Belle Vernon. Each town awarded prizes for the best costume by age group, and Lana wanted to win one in hers.

Lana was a cutie, with blond hair, sky blue eyes, and rosy cheeks, and she loved dressing up. For the 1948 parades her parents dressed her as a princess. She wore a white, tulle-lined gown with gold sparkles, a cardboard crown, a bright red cape tied in front with a red ribbon, white

FIG. 21.1. Lana Mills in a princess costume made by her mother. Lana marched with her father, dressed as a clown, in the 1948 Halloween parade in Donora. She carried a homemade wand with a lightbulb on top and batteries in the handle. Courtesy of Lana Vrabel.

gloves, and saddle shoes. She carried in her hand a wand with a lighted tip that her father had cobbled together. She could turn it on and off with a toggle switch on the handle. The costume was completely homemade. "Back then," recalled Lana, "you didn't go to a store and buy your costume. Everybody made them."

Little Lana made quite an impression that day in Donora's parade. "When you came down the street," Grace told her daughter later, "we couldn't see the marchers until they were really close because it was so dark. All we could see was the light from your wand!"

So impressed were the judges with Lana's costume that they awarded her first prize in her age group. What role the lighted wand played in their selection is unknown, though it couldn't have hurt.

A large number of area residents had managed to find their way downtown for the parade. Most of them tied handkerchiefs over their mouth and nose. Of the hundreds of onlookers few were able to see much more than "shadows marching by." It was as if the people riding in cars, trucks, and floats were foggy apparitions, actors behind a frosted scrim.

———————————

John William Volk rode one of the firetrucks in the parade. He was the fire chief and had ridden in parades many times before. Volk was close to six feet tall and skinny, just 135 pounds. Born October 21, 1901, the blond, blue-eyed Volk had just turned forty-seven when he appeared in the 1948 Halloween parade. He had a square jaw, was missing a front tooth, and wore wire-framed glasses perched on his wide nose. There was a kindness to his mien and a gentleness that would have put distressed homeowners quickly at ease.

Donora's mostly volunteer department had been incorporated in 1901, naming D. F. Millison as its first permanent fire chief. The company had grown over the years and in 1948 consisted of about thirty volunteers and two full-timers: Volk and his assistant fire chief, Russell Davis. After the parade both men returned to the station rather disappointed. "As a rule I like a parade," said Volk. "We've got some nice equipment here, and I don't mind showing it off. But I didn't get much pleasure out of that one. Nobody could see us, hardly, and we couldn't see them. That fog was black as a derby hat. It had us all coughing."

The fog bothered nearly everyone, even children, most of whom were wearing cloth coverings on their face. The kids didn't really care, though, they just wanted to enjoy a normal Halloween. Patricia Dugan was a Halloween-loving twelve-year-old who wanted to go trick-or-treating with her friend, and she didn't care how awful the fog was. "We were going to go trick-or-treating no matter what," said Patricia, "because it was trick-or-treating time. We tied stuff around our mouths. We got about probably less than a block from our house before we had to turn around and go home. We couldn't see. Our eyes were burning and everything else. You couldn't see five feet in front of you. We turned around and, disgusted, we went home."

Edie Jericho and a few of her friends from Webster walked over the Donora-Webster Bridge to watch Donora's Halloween parade, at least what the twelve-year-olds could see of it. "You could see the people who were close to you," Edie recalled, "but you couldn't really see across the street and tell who was on the other side."

After the parade Edie walked home and told her mother that she was going trick-or-treating with her friends. "Oh, no, you're not," her mother said, "it's bad outside." Edie insisted, as only a determined twelve-year-old can, saying, "I *have* to go trick-or-treating! I'm going to Olive's!" Olive Lorenzo, Edie said, "made the most delicious candy apples you ever ate in your life." Her mother could not refuse the young lady's request and let her daughter walk the short way to Olive's.

So thick was the fog that when the parade ended, the crowd rapidly dispersed. "You really couldn't see a thing," recalled Georgette Chambon. "I was glad to get back in the house. I guess everybody was. The minute it was over, everybody scattered. They just vanished. In two minutes there wasn't a soul left on the street. It was as quiet as midnight."

Around the time the parade started Bernardo Di Sanza was taking a turn for the worse. The Adrenalin injection he received had not worked as well as Rongaus had hoped. Di Sanza's lips began to turn blue. His hands shook, partly from the Adrenalin and partly from the lack of oxygen in his body. He felt bloated, his chest hurt, his vision was fuzzy, and his eyeballs seemed to be popping out of their sockets.

Di Sanza slept off and on throughout the night. His family placed three pillows behind his back and head to keep his torso elevated. That position proved the only way he could breathe more easily, and it relieved his chest pain as well.

Had Rongaus or any another physician seen Di Sanza at that point they surely would have called for an ambulance to take him to the hospital. Whether there would have been an ambulance available is another question. Ambulances had been ferrying fog victims to the hospital all evening, one after another. But no physician came to the Di Sanza home; all of them were out making house calls. They couldn't get to everyone who needed them. It just wasn't possible. Rongaus that evening was working with Richard Lawson, owner of a funeral home, to make house calls. Lawson drove a Board of Health ambulance as Rongaus walked in front, leading the way.

"Before the evening was half over," Rongaus told a reporter later, "a sort of panic seized the town. I would go into a home to give a treatment. When I would leave, [neighbors] would be waiting on their front porches to have me treat somebody in their house. It got to the point where I was going almost from door to door. There wasn't time to get everybody to the hospital. I would give one of the patients a treatment and go on to make more calls, but the treatments were only effective a few hours, and before I could return some of my patients died."

22

THE TOWN
REACTS

CHIEF VOLK AND HIS YOUNG ASSISTANT, RUSSELL DAVIS, HAD JUST SAT down for a cup of coffee when the phone rang. It was perhaps 8:30 p.m. or so. The men had just returned to the fire house from the parade, as had several volunteers who marched alongside the trucks. The chief and his assistant said their good-nights to the volunteers and began to rest for a few minutes before heading home. When the phone rang they both looked at it, stunned into silence.

A fire on such a foggy night? Volk thought. *That could be real mean.* He dreaded answering it, but when he did he found it wasn't a fire at all. "It was a fellow up the street," said Volk, "and the fog had got him. He said he was choking to death." The man said he hadn't been able to reach any doctor and that he needed oxygen. Davis watched his boss's mouth drop open and knew that whatever the caller said couldn't be good.

Volk hung up and stood there, trying to think what to do. He believed that the man was telling the truth, that it wasn't some Halloween prank. He sounded "half-dead" already. Firefighters generally didn't go around treating the sick; doctors were supposed to do that. "But what the hell," Volk said, "you can't let a man die!"

So he told Davis to get there and do whatever he could. Harry Russell Davis looked every inch the prototypical fireman. Five-ten, 170 or so pounds, and handsome, with striking gray eyes and a thick brown mane, Russell had been a fireman for several years. His interest might have stemmed, at least in part, from a traumatic childhood event. Born September 17, 1912, Russell was just eight when his sixteen-year-old sister Kathryn suffered a horrific accident. Kathryn was burning some paper in the family's backyard when her clothing caught fire. Panicked, she "dashed madly up the street, fanning the flames and causing them to spread." She sustained such severe burns that doctors thought she would die from her injuries. She survived, however, and became a journalist in New York City. She died eighteen years later, taking her life in the apartment she shared with her brother Robert.

Russell was not yet finished with family tragedies, however. Just two years after Kathryn burned herself almost to death Davis's father, Harry Clay Davis, perished in a train accident. The elder Davis was a yard foreman for Donora Southern Railroad. On May 1, 1922, he was trying to couple two cars when he was caught between them and "horribly squeezed about the abdomen." He died at midnight that night. He was thirty-nine years old.

Two such dreadful events before he was yet ten years old must have affected Davis to his core. So when Chief Volk told him to help the "half-dead" man, Davis jumped to the rescue. If he couldn't have saved his sister and father, maybe he could save this neighbor.

Davis quickly grabbed a portable oxygen unit, jumped into his car, and took off. The unit consisted of a dark green tank filled with forty cubic feet of compressed oxygen. A valve and pressure gauge was clamped on the top of the tank and led to a thick, black rubber corrugated hose about three feet long. A dark green rubber face mask dangled from the end of the hose.

After tending to the first caller, Davis learned that other calls had come in. "That guy was only the first," said Davis. "From then on, it was one emergency call after another. I didn't get to bed until Sunday. Neither did John. I don't know how many calls we had."

Davis traveled from house to house through the intense fog all night. He found that the only way he could drive was to drive on the left side of the road, stick his head out of the window, and listen for the car's tires scraping along the sidewalk.

At some point during the night the car refused to idle when stopped. The air filter in Davis's car was so clogged with dust and soot that if the car stopped for even a moment the engine would shut off. Combustible engines can't run without air. "I'd take my foot off the accelerator," he recalled, "and the engine would stall. There just wasn't any oxygen in the air. I don't know how I kept breathing. I don't know how anybody did."

All night long Davis visited people struggling to breathe. Some lay on their beds, some on the floor, some even in the basement, with their head wedged inside the furnace draft, trying to find more air. Davis would sling a sheet or blanket over the person, stick the inhalator's mask underneath, and crack open the valve to release a light stream of oxygen. He let the gas flow inside this makeshift oxygen tent for about fifteen minutes per patient. That short burst of oxygen quickly improved the person's breathing. "By God, that rallied them!" said Davis.

In many cases, and possibly all, those fifteen minutes of oxygen provided enough of a respite that the person survived through the weekend. No records exist of how many people Davis visited that weekend, where they lived, or who they were, but it seems reasonable to conclude that, of the perhaps several dozen he treated, the overwhelming majority survived.

Davis soon started to become short of breath himself, though, and he began to cough. He didn't want to use any of the oxygen for himself, so he just kept going. When he ran out of oxygen he drove back to the fire station for more tanks, and when he got there he drank "a little shot of whiskey," Davis said. "That seemed to help. It eased my throat."

Volk continued to receive emergency calls after Davis left the station that first time. He needed help, so he rang Bill Schempp at home. Schempp was one of his best volunteers, if not the best, and he needed him. Now.

If there was a fire, Volk would have alerted all available volunteers with a system the department had been using for several years. Each firefighter was given a small box with a bell and light bulb built into it. When a fire occurred a signal would be sent from the fire station to the box. The bell would ring a certain number of times and the light bulb would flash a certain number of times, the combination of which indicated the approximate location of the fire. The volunteers each had a map, perhaps a chart or table, that indicated a particular area. For instance, the blocks between Meldon and Thompson Avenues and bordered by First and Second Streets might have been one area. That area would have had its own set of rings and flashes.

Volk couldn't do that for the fog, though. The firefighters wouldn't know which home in the area housed the sick person. So Volk had to use the phone to communicate with his team. It is possible that Volk used portable two-way radios, or walkie-talkies, to communicate. Walkie-talkies were used extensively in World War II for command posts to communicate with soldiers at the front. After the war the devices grew in popularity among police and fire departments and were even being sold as toys.

Gladys Schempp wasn't home when the phone rang. She was at a party for a local troop of Brownies. The girls had marched in the parade and were now having a few treats. Bill heard the phone ring and walked into the kitchen to answer it.

Volk told Bill about the calls he had been receiving and that he needed volunteers to visit homes. Schempp was a helper, someone who would go out of his way to assist anyone, anytime. At five-seven, and with a wide smile always at the ready, Schempp was engaging, inviting, and as honest as the sound of an infant's giggle. Born January 27, 1917, four months before the United States declared war on Germany, Schempp was twenty-five when he joined the army, in 1942. He was assigned to a horse cavalry unit in Texas. While there he learned how to care for horses, ride them in formation on a variety of terrains, and even jump them on an obstacle course.

Shortly before Schempp's platoon was sent to the South Pacific, he injured his back. According to his daughter, Annie, a group of men officers and a woman magazine writer had been drinking heavily at the officer's club and decided that they wanted to watch some horses being put through their paces. The officers told Schempp, a corporal at the time, along with a few cavalrymen, to put on an impromptu jumping exhibition,

FIG. 22.1. Gladys and Bill Schempp. Courtesy of Annie Schempp.

but Schempp balked. "We can't jump without warming up the horses," he told the officers.

Just as in humans, a warm-up increases the heart rate and boosts circulation, which in turn increase blood flow to muscles and helps loosen joints. For horses specifically, a warm-up period would have allowed the animals, unaccustomed to late-night activity, to recognize that they were being put to work. Without even a brief warm-up, the horses might have been confused about their role. They were certainly ill prepared for jumping on command, and consequently they and their riders faced a significantly higher risk of injury. Schempp tried to explain all that but was quickly overruled and ordered to put on a show.

Schempp was injured while jumping that night. Exactly how the injury happened is unclear, but it seems he was thrown from the saddle when his horse unexpectedly stopped at a hurdle or perhaps jerked to one side of it, common causes of rider injuries on obstacle courses even today. Regardless, Schempp suffered severe damage to his back. "His spine was crushed," said daughter Annie. "They took a piece of bone out of his leg and put it in his back. He was in a full body cast for years."

Discharged in 1945, Schempp never saw combat. He returned home in constant pain and under the watchful eye of a vigilant wife. Said Annie, "My father always attributed his life to her because she wouldn't let him just sit around the house. So he joined the fire department, and it saved his life." Even though Schempp was in constant pain, he did everything other firefighters did, including carrying people from burning buildings and hauling around heavy oxygen tanks to treat residents with smoke inhalation. He dealt with his pain quietly, secretly. Only Glady knew how much he hurt, which is how he wanted it.

So when Volk called Schempp that night asking for help, Bill couldn't possibly have refused, no matter how tired he might have been or how much his back might have hurt. He simply wasn't made that way.

He grabbed his turnout coat, a heavy, three-quarter-length, mustard brown waxed canvas jacket, with snaps on the inner flap and four large metal hook-and-eye closures on the outside. He slid his feet into a pair of black rubber boots and snugged his Donora Fire Department helmet onto his head. He strapped on his back the oxygen tank he kept at home, the green one, labeled TO BE FILLED WITH COMPRESSED OXYGEN ONLY, and wrote Gladys a note about where he was going. Then he walked out the back door and turned left onto Thompson Avenue, toward the fire station. He didn't have far to walk: the Schempps lived at 625 Thompson, and the fire station was two blocks south, at 422 Thompson.

A painting contractor in town, Jim Glaros, also responded to the call. Glaros, at fifty-one, was one of the older firefighters in town. No matter. He grabbed his own oxygen tank, and he too headed to the fire station, though he had a considerably longer walk, just under a mile.

Volk had been fielding a stream of calls about people having trouble breathing when Schempp arrived. Volk gave him a list of homes to visit, and another list to Glaros when he walked through the door later. Then each man turned and walked out of the fire station, into the stygian night.

23

DEATH BEGINS
ITS ASSAULT

IT WAS AROUND MIDNIGHT ON FRIDAY WHEN EDWARD ROTH PULLED HIS
car to the side of the road and shut the engine off. The fog had become
so thick and so dark that driving had become impossible. He grabbed his
medical bag and left the car where it was. Then he started walking, feeling
his way along the sidewalk.

It seemed every physician in town was doing the same. American Steel
& Wire's Dr. Hannigan was making house calls and sometimes visited the
same people another physician had already seen. "We all had practically
the same calls," Hannigan remembered, "Some people called every doctor
in town. It was pretty discouraging to finally get someplace and drag your-
self up the steps, and then be told that Dr. So-and-So had just been there.
Not that I blame them, though. Far from it. There were a couple of times
when I was about ready to call for help myself. Frankly, I don't know how
any of us doctors managed to hold out and keep going that night."

Ralph Koehler had been wending his way from house to house as well
and had finally reached his breaking point. "I had to go home," Koehler

said. "God knows I didn't want to. I had hardly made a dent in my list. Every time I called home or the Physicians' Exchange, it doubled, but my heart gave out. I couldn't go on any longer without some rest." As he dropped into bed at home he heard his wife answering the phone. Then he fell into a deep sleep.

Working at the wire mill had taken its toll on Ivan Ceh. He was sixty-nine now and had suffered from an inflamed heart, a condition called myocarditis, for a number of years. He made it through the first few days of the smog, but by one o'clock Saturday morning he had been coughing and short of breath for several hours and was getting worse.

Ceh (pronounced *tseh*) was just twenty-three when he first touched American soil. He was born in Rijeka, Croatia, an ancient city perched along the northern Adriatic Sea. Rijeka is lined with dazzling neoclassical palaces and has served as a major seaport for centuries. Ceh left from Le Havre, France, on September 7, 1902, on the steamship *La Lorraine* and landed in New York City six days later. He found his way to Donora and landed a job as a cooper at American Steel & Wire.

A single man, Ceh lived in the rear of the house at 634 Fifth Street, about halfway up the hill and around the corner from Roth's office. Ceh was a leader in the town's Croatian community and in 1915 helped to establish a local branch of the National Croatian Society. The so-called Cro Club met at the corner of Fifth and Hickory for many years. The group grew in size and in 1951 opened a new home, an impressive orange-brick structure at the corner of Fourth and Castner.

Ceh's apartment, like almost every other home and apartment in town that weekend, had been filling with smog. It came through cracks in the foundation, fissures along windowpanes, and crannies under exterior doors. The smoke wasn't as thick inside as it was outside, but it burned the eyes and clogged the lungs just the same.

His breathing worsened as the night wore on. Although no records exist of his last few hours, anyone with myocarditis living in those smoky conditions would have experienced severely labored breathing. The heart, already vulnerable from chronic inflammation, would have beat faster and faster to push whatever oxygenated blood was available to the brain, lungs, liver, and kidneys, the body's most vital organs. Ceh would have become progressively weaker, his heart rate and breathing becoming ever faster but less effective.

Without a sufficient supply of oxygen reaching his brain, Ceh's thinking would have clouded. He might have had chest pain, and his feet almost certainly ballooned from fluid accumulating in the tissues. He would have focused entirely on breathing, unable to think of much else but pulling in enough air to feed his oxygen-starved body. In all likelihood he knew, at least on some level, that his life might be approaching its end. Those thoughts and feelings would have aroused in him a sense of dread, of absolute terror, that only those in the grip of their own mortality could fathom.

In the last hour, perhaps less, his breathing would have slowed and shallowed. His heart rate would have dropped. His heart and lungs failing, he would have lost awareness of his surroundings, finally slipping into unconsciousness. At the last his breathing would have become irregular, his breaths coming in great gasps, until finally a last breath would have oozed out of his mouth, as if sticking a pin in an old, soft balloon. Then, stillness.

Someone with Ceh at the end, possibly Viola Vrankovich, noted a whitish foam around the man's lips. Vrankovich was the widow of Marko Vrankovich, a Donora baker who had died of pneumonia years before. Viola and Ceh lived in the same rear apartment. If it was indeed Viola who had noted the white foam, Ceh was at least not alone in those final moments. It was 1:30 a.m. when the retired barrel-maker breathed his last.

Thirty minutes after Ceh died someone called the R. R. Schwerha Funeral Home. As it turned out, Ceh's body was only the first of ten to be collected that weekend by Schwerha's mortuary.

Rudolph Robert Schwerha was a familiar figure in Donora. A longtime member of the Donora School Board and a prominent member of the Saint Dominic's Men's Club, Schwerha was born in 1900 in Braddock, Pennsylvania, just south of the Carrie blast furnaces. He worked as a milkman after high school, delivering milk and cream in glass bottles sealed with a waxed foil cap. He attended Eckels College of Mortuary Science in downtown Philadelphia, a small school occupying a few row homes along the 1900 block of Arch Street. After graduating in 1921 Schwerha returned to Donora and set up a funeral home at 546 Thompson.

Schwerha was a dapper businessman. He kept his thick, dark brown hair closely trimmed over his ears and let it go longer on top. A slender mustache perched atop his upper lip lent an air of professionalism to the man. He was someone a grieving family could trust, and for a mortician trustworthiness was a critical asset.

It was 1:30 a.m. when Schwerha sent his driver to pick up Ivan Ceh, just a few streets away from the funeral home. The call about Jane Kirkwood came thirty minutes later, just as Schwerha's driver was returning.

Jane "Jeanie" Kirkwood was just a month shy of her seventy-seventh birthday when the fog hit. Born in Cambusnethan, a suburb of Wishaw, Scotland, Jeanie and her husband, Thomas, along with their daughter, Mary, had left Scotland by steamer in late 1911, landing in Nova Scotia. Some time after that the family moved to Donora, where Thomas found work as an engineer at American Steel & Wire. Thomas died in 1935, after which Jeanie moved in with her daughter, whom everyone called "Minnie," and her husband, John B. Clark. They lived in a spacious home at 121 Ida Avenue, a quiet street lined with sycamore trees in the middle of Cement City.

Jeanie awoke around four Wednesday morning with a dry cough. She was short of breath and just wasn't feeling well. The fog always made her feel this way. She had suffered from heart failure over the years, so anytime the fog hit, her heart struggled to keep up. Kirkwood felt increasingly sick as the morning wore on. Someone, probably Minnie, called for a doctor. One of the town's general-practice physicians arrived later that morning and injected Jeanie with morphine or Adrenalin. It didn't help, and the doctor was called again that night. He gave Jeanie another injection, hoping the second one would be enough to calm her breathing and let her heart rest.

It wasn't. She couldn't sleep at all. Her breathing was so labored that she couldn't lie flat, so she either sat upright in a chair or used pillows to prop her torso in bed. Nothing helped. Her heart was too weak to continue, and she died at two o'clock Saturday morning, becoming the second victim of the smog.

At the same time Jeanie Kirkwood was drawing her last breath, a retired coal miner named Peter Starcovich was also struggling through his last moments on earth. Starcovich lived in a house across the river, north of the Zinc Works, in a highly rural area known as Wilko Hill. Starcovich had come to the United States from Yugoslavia as a youth. He moved to Donora in 1927 and retired three years later "because of an injury." He developed a dry cough and dyspnea early Friday morning. He vomited several times throughout the afternoon, and his breathing continued to deteriorate. Peter's wife, Julia, might have called for a doctor, but if so, no physician ever came. Starcovich died at 2:30 a.m., just thirty minutes after Kirkwood.

Schwerha knew nothing about the death of Jeanie Kirkwood; he was too busy trying to get Starcovich's body back to the funeral home without running off the road.

Up on the hill above Webster, in the hamlet of Fellsburg, a fifty-eight-year-old widow was struggling for her life. Ida Orr had been born in Alabama and was the widow of John "June" Orr, a farmer. Ida had fallen ill Friday morning, awakening at 6:30 a.m. with dyspnea. Factory smoke had bothered her many times before, but not like this. She developed a dry cough and became progressively more short of breath. No doctor visited her; she lived three or more miles away from the center of Donora, and even if she or her family had been able to reach a physician, it would have been enormously difficult for any of them to have trekked all that way in the dense fog. Ida weakened throughout the day, her body finally giving out at 3:00 a.m. Saturday morning.

Not quite two hours later, at 4:55 a.m., John Cunningham fell victim. Cunningham had been born in Martinsburg, West Virginia, three days after Christmas 1884, possibly 1883. He had lived in Donora for thirty years, working his way up from a laborer to an experienced rod bender at American Steel & Wire's steel mill. Cunningham and his wife, Harriet Luella, made their home at 322 Tenth Street, just as the road bends south past Heslep Avenue. Luella, who died at sixty-four from heart and kidney disease, had been gone just over a year and a half when the fog felled her husband.

Cunningham had just started work at Rod Mill 3 that Friday morning when he started feeling as if he couldn't catch a breath. Co-workers helped him to the plant's medical department at 3:00. A doctor there gave him a prescription for a few pills, possibly theophylline, a drug used to open the airways in asthmatic patients. Why he didn't receive an Adrenalin injection is unclear; he had a prior history of receiving Adrenalin and morphine injections for asthma attacks. Perhaps the mill physicians considered the case too mild for Adrenalin. Perhaps the department had run out by then. Whatever the reason, Cunningham's breathing continued to worsen at home, even though he had been taking the drug prescribed by the mill physician.

He couldn't lie down in bed without his dyspnea intensifying. He fell asleep around seven o'clock, but his sleep was undoubtedly fitful. His breathing became progressively shallower, his heart rate slowing, until he finally succumbed.

Five deaths in less than three and a half hours, three in Donora proper and two more on the hill across the river. And hardly anyone knew why.

24

"OH, HELEN, MY DAD JUST DIED!"

BERNARDO DI SANZA, THE HAPPY-GO-LUCKY RAILROAD MAN FROM Rocco Pia, Italy, the hale and hearty track foreman who had retired from Donora Southern Railroad not even a year before, continued his battle to breathe. He had barely slept, nodding off and on all night as his wife, Liberata, stayed by his side.

Early Saturday morning he tried to climb out of bed, but he was so weak that he collapsed on the floor. Liberata must have been terrified to see her otherwise healthy husband lying in a heap, his hands and feet flailing in a vain attempt to arise. She surely saw the panic in Bernardo's eyes, now glassy from the strain. How utterly desperate she must have felt, powerless to help.

Bernardo's strength carried him through the morning, his will to live so strong that he persisted when others would have perished. By midmorning he began to babble. His oxygen-starved brain could manage only incoherent mumbles, something about his life in Italy a half century ago. An hour later he uttered his first lucid sentence in hours: "It's raining outside, and I can see the zinc plant." It wasn't raining, and from his bed he couldn't see much of anything outside. He was hallucinating, a common occurrence in people with markedly low levels of oxygen in their blood.

Di Sanza would soon rally, his mind would clear, and he would ask for food.

Andrew Odelga had been sick since a brief episode of dyspnea Wednesday afternoon. Born in Zvolenska, a tiny village in central Czechoslovakia, Odelga was twenty-four when he immigrated to the United States, arriving in New York Harbor in 1903 aboard the two-year-old Norwegian steamer SS *Moltke*. Odelga was sixty-nine and had retired from his position as a welder at American Steel & Wire three years before.

Like Di Sanza, Odelga had been healthy, bothered only slightly by an occasional fog. So when he became somewhat short of breath that Wednesday, he almost certainly thought nothing of it. At five-nine and 230 pounds Odelga wasn't terribly fit and might have already developed diabetes or high blood pressure, common ailments in overweight, nearly seventy-year-old individuals. If so, no record exists of those or other underlying conditions. The dyspnea didn't last long, and Odelga felt fine until Friday afternoon, when he decided to sit on the front porch of his home at Eighth and Van and watch the day go by. The fog was thick, dark, and stinging. Why he chose to sit outside will never be known, but whatever the reason, he became short of breath again, and this time there would be no letup.

He or someone in his family was able to reach a physician, who visited the home later in the afternoon and gave Odelga an Adrenalin injection and also a pill, probably theophylline. The medications seemed to help for a time, but that evening around 7:30 Odelga's dyspnea worsened, and his abdomen began to bloat. Patients with dyspnea frequently take short, shallow breaths, and in doing so they can swallow air. Air entering the stomach can then pass into the intestines, causing bloating and an uncomfortable feeling of fullness.

Odelga's breathing worsened all night. His body weakened, and around the time Bernardo Di Sanza was falling out of bed, Odelga was passing away.

Ralph Koehler had been so exhausted Friday night that it seemed as if he had just laid his head on the pillow when it was time to get up. He dressed as quickly as he could, and then rushed downstairs to quaff a cup of coffee before heading to the office. The phone rang before he finished. It was Helen Stack, updating him about patients to be seen, one of whom was the father of one of Helen's best friends, Dorothy Hollowiti. Stack asked Koehler to see Dorothy's dad, Ignace, as soon as he could. Ignace Hollowiti had been born in 1883 in the Kingdom of Galicia and Lodomeria, a region that straddled the border of today's Poland and Hungary. He had been working at the wire mill as a fireman and laborer for the last thirty-seven years.

After speaking with her boss, Stack dialed Dorothy's phone to tell her that Dr. Koehler would be there shortly. It was 6:45 a.m.

"Dorothy, it's Helen. The doctor is on his way over now."

"Oh, Helen," cried Dorothy, "my dad just died. He's dead!"

The words struck Helen speechless. She searched for calming words, something that would ease her friend's grief. "I don't remember what I said," she recalled later. "I was stunned. I suppose I said what people say. I must have. But all I could think was, *My gosh, if people are dying—why, this is tragic! Nothing like this has ever happened before!*"

People were indeed dying, and not just in Donora.

Annie Troy and her husband, James, lived in a small house in Webster. Their lawn—mostly dirt and rocks—was ringed by a bare wood picket fence. If it had ever been painted, the paint had long ago been eaten away by chemicals from zinc smelter smoke from across the river. James worked at the smelter; Annie kept house for him and their four children.

Emma Hobbs, a Black woman, lived next door to the Troys, a White family, in a tiny, time-battered house. She was short, barely five feet tall, and plump. Hobbs always wore a sweet, caring smile on her weathered face. "My mother loved Emma," recalled one of Annie's daughters, Edie Jericho. "Emma would come to the fence and holler, 'Miss Annie, Miss Annie!' Emma was from the Old South, and she wouldn't walk through the fence gate and knock on the door; she'd just holler. My mother would go outside, and they would talk over the fence. They would talk for a long time. I think a couple of times my mother said, 'Come and sit on the porch, Emma.' But she never did."

Over the years Hobbs and Troy had forged a deep friendship, and each knew she could count on the other. "One time, when my mother was pregnant with my brother," Jericho said, "my sister was just learning to ride a bike. Her bike didn't have a chain guard, and her pants got caught and she fell. My mother heard her screaming and took off running. She tried to jump over this little ditch, but she fell when she landed and cut her leg open on a piece of glass. Emma was the first one out there taking care of her."

Hobbs had a heart condition and was often bothered by fog in the valley. On Friday morning she developed dyspnea and cough and felt as if she was choking. Bill Rongaus visited around midnight. He gave her an injection, but it didn't help. Rongaus told Hobbs and John Bazure, a man who lived with Emma, "You should get out of town. That stuff is coming over here, and every time it blows you know it hurts you." They had nowhere else to go, though, so they stayed in the house, hoping for a miracle.

Around 7:30 Saturday morning, Annie Troy heard a voice calling out. It was Bazure standing at the fence, screaming, "Miss Annie! Miss Annie!"

Annie rushed outside and asked, "What's the matter?"

"Hurry up!" he told Annie, "Something's really wrong with my Emma!"

Annie ran next door, but it was too late. Her dear friend was gone. Emma was just fifty-five years old.

Perry Stevens, a year younger than Hobbs, died about fifteen minutes later. Stevens was fairly new to Donora, having moved there six years before. He was short, five foot five, and just under 160 pounds. Born in Birmingham, Alabama, Perry had lived here and there throughout the region, mostly staying in boardinghouses. He worked for a number of years as a coal miner near South Fayette, Pennsylvania, a few miles southwest of Pittsburgh. Stevens might have worked coal mines around Donora as well, but when the fog hit he was unemployed and receiving federal welfare assistance.

Perry developed a heart condition in 1944 called auricular fibrillation ("atrial fibrillation" in today's parlance), a condition which left him with an irregular pulse and a heightened tendency to have a heart attack or stroke. He had been treated then with digitalis and might still have been taking it in late 1948. Perry also suffered from high blood pressure and occasional heart failure, so his cardiovascular system was already weak when the October smog rolled in.

Perry's illness began about an hour before daybreak on Thursday with a brief bout of dyspnea and cough. He felt well enough later that morning to walk the six blocks to downtown, but the walk proved too much for him, and a friend had to drive him home. He survived through Friday night, but his heart and lungs gave out early Saturday morning.

It was not yet eight o'clock in the morning and already nine people were dead. Hardly anyone knew about it, perhaps a few dozen individuals in the whole valley, and they were mostly the families of the victims, the physicians and firefighters working their way around town to care for the ill, and the undertakers who collected the remains of the dead. Almost everyone else continued on with their day, oblivious to the tumult around them.

———

Not even the burgess knew anything unusual was happening. The position of burgess, or mayor, at the time was mostly ceremonial. August Chambon certainly couldn't make a living from it; his living came from the moving and storage business he owned. He was in Pittsburgh Saturday morning and wouldn't return until around 2:00 p.m. His wife had already told him over the phone about his mother and a few others around town being sick, but he didn't worry. "I just thought they were like Mother," Chambon said, "old people that were always bothered by fog. Jesus, in a town like this you've got to expect fog, it's natural. At least that's what I thought then."

Unable to reach the burgess, John Elco and Cora Vernon decided to set up an emergency aid station. Elco was an electrician from Forward and a longtime member of the Ernest E. Jobes Post No. 212 of the American Legion in Donora. Vernon was a registered nurse and director of the local Red Cross. Vernon learned that Charleroi-Monessen Hospital, the hospital closest to town, was completely filled with valley residents sick from the smog. "We can't take another patient," said Dr. James Lau, super-

FIG. 24.1. Residents gather in front of the community center during the fiftieth anniversary of the town's founding. A temporary medical aid station was installed here during the smog. The entrance was through the first-floor porch, between the columns. The ambulance entrance was probably around the corner, past the Public Library sign at lower left. Courtesy of Donora Historical Society.

intendent of the hospital. "All day long I've been shuttling back and forth between the hospital and Pittsburgh getting extra equipment. We've got eleven oxygen tents in operation tonight, compared to our usual three. So far, however, we've been lucky. None of the deaths have occurred at the hospital."

Elco and Vernon thought an aid station in town could provide at least some assistance for the sick. They decided to turn the community center into a temporary hospital. The community center occupied the first floor on the Seventh Avenue side of the old Donora Hotel on McKean. The basement of the Center would serve as a temporary morgue; it had a separate entrance that would allow for the unobtrusive conveyance of a body from an ambulance or a hearse. The basement typically hosted meetings of the local Brownie, Girl Scout, Cub Scout, and Boy Scout troops. Tonight it would host the bodies of smog victims.

The town's doctors and firefighters had been scurrying around town since Friday night, delivering whatever care they could for what seemed to them a never-ending stream of patients. Rongaus, for instance, was about seventy-five calls behind. "There were thirty calls waiting for me at the Board of Health, about thirty at the hospital, and about fifteen calls at my home. I just couldn't keep up with them."

Firefighters, too, were working around the clock. Besides Bill Schempp, Russell Davis, and Jim Glaros, the other volunteer firefighter known to have helped was young Paul Brown, a tall, lanky twenty-one-year-old laborer at the steel mill. At six feet four and 160 pounds, Brown's teammates on his high school football team called him "Beanie Davis." Even the chief made calls to deliver oxygen. "I found people laying in bed and laying on the floor," said John Volk. "Some of them didn't give a damn whether they died or not."

Communication among all of those physicians, firefighters, and the patients who needed them proved difficult and haphazard. There were no cell phones, no internet, no centralized communication system of any kind other than two telephone switchboards, one in the mill complex for employees and one for the town, on the second floor of the building at Sixth and McKean.

Miriam Alice Uhriniak, known by everyone as Alice, started working at the town's Bell Telephone switchboard after graduating Donora High School in 1945. Alice was an attractive, dark-haired twenty-one-year-old when the smog hit. According to her yearbook she was "tardy on occasions," but she was on time that Saturday, arriving to work at 7:00 a.m. Alice walked into a flurry of activity. The operators were all talking on their headsets and switching cables around like fast-chess players swapping pawns in New York's Washington Square Park. "Hurry up, and get your headset on," one of the night operators told her. "Everybody is dying!"

People were calling for doctors, doctors were calling their office, and offices were calling Charleroi-Monessen for ambulances to pick up patients. "The phone lines were absolutely flooded," said Donora historian Charlton. "It was just absolute and total chaos."

Physicians also made use of their answering service, called the Physicians Exchange, to learn who needed their care. The Physicians Exchange started in December 1938 by one of town's nurses, the same person who answered calls that weekend: Elizabeth Ostrander. Ostrander was the daughter of Eleisa Abbey Ostrander, one of the town's original residents, who had moved there in 1901. Eleisa opened the town's first ice cream parlor and had twelve children, one of whom, Mae, became a well-known schoolteacher in town, and another, Elizabeth, who not only operated the town's first and apparently only physicians call service but would later serve as secretary of the town's Board of Health. Elizabeth ran the service much like today's answering services, taking calls from patients, recording their requests, and passing the requests along to the appropriate physicians. She operated out of the Ostrander family home at 402 Thompson, just steps away from the fire department.

Davis, Schempp, and the other firefighters went through oxygen tanks like pots of morning coffee in a busy office. Each tank at the station held 255 liters of oxygen. How long each tank lasted depended on how long each firefighter let the oxygen flow and at what rate. Davis's blanket-over-the-head setup, for instance, would have needed a new tank about every six patients. He had a car, so he could store extra tanks and then drive back to the fire station when he needed more.

Bill Schempp decided it was too foggy to drive his car, so he walked around town, feeling his way along buildings, fences, signposts, anything he came across. There were times he had to crawl along the road to feel his way forward. "If you chewed hard enough, you could swallow it," said Schempp about the thickness of the fog. "It almost got to the point where it was claustrophobic, it was so dense and thick. You had to get right up to the door and guess where you were."

Once inside, Schempp would drop his gear near the patient, fit a thick rubber mask over their nose and mouth, and turn on the oxygen. He let the gas flow for ten or fifteen seconds, delivering what he called a "shot of oxygen." He usually gave one or two shots, sometimes three, over a period of five or ten minutes. With such short bursts of oxygen Schempp's tank would have lasted considerably longer than one of Davis's tanks, but still, Schempp would have had to change tanks about every fifteen or so patients. And there were thousands of people sick from the smog.

Davis and Schempp faced an excruciating decision in every house. At some point each man would have to shut the oxygen off, collect his gear, and move to the next house, to the next person clamoring for help. Physicians visiting homes fared better in that regard. After all, they knew that whatever medications they administered would work long after they left the house. People understood that doctors were busy people, that they would check them over and give them some kind of treatment, even if it was just a prescription, and then leave.

Not so with firefighters. They stayed until the fire was extinguished, the scared little kitty was safely back on the ground, the car accident fully cleared. So it was that Schempp, Davis, Glaros, Brown, and Volk—men who had fought house fires, transported the sick and injured to local hospitals, comforted those who had lost their home or loved ones—were forced to decide how much oxygen to give each patient and when to leave. They had to say over and over, "No, I'm sorry. I have to go." They had to look into the eyes of every one of those critically ill people, of their family members and friends, and turn away. They had to listen to the family beg them to give their loved one more oxygen. "I'm dying," the patients would say, "and you're taking my air from me."

The men had to listen to those gut-wrenching pleas and then walk away knowing they might never see those people alive again. Such torment. Said Schempp, "It almost broke my heart to leave."

Burgess Chambon found a message waiting for him when he arrived home around 2:30 p.m. It was from Elco, urging him to meet at the Borough Building as soon as he read the note. "I wondered what the hell," said Chambon, "but I went right over. It isn't like John to get excited over nothing. The fog didn't even enter my mind, so I was astonished when John told me that the fog was causing sickness all over town. I was just about floored. That's a fact. Because I felt fine myself. I was hardly even coughing much."

By that time Cora Vernon had already alerted the local Board of Health and the Pennsylvania Department of Health to the crisis. The agencies immediately requested assistance from the US Public Health Service, one of the oldest US government departments. The USPHS, a military agency, had been formed in 1798 when President John Adams signed into law the

Act for the Relief of Sick and Disabled Seamen. The act led to the construction and staffing of several marine hospitals along major US waterways. By 1944 the USPHS Commission Corps, as it had become known, was authorized not only to prevent disease but also to research diseases, sanitation, water supplies, and sewage disposal. The corps had grown by 1948 to include nearly three thousand physicians, nurses, dietitians, physical therapists, and other healthcare professionals in its ranks, all serving at the direction of a surgeon general.

The request from the state health department reached the USPHS Saturday morning, but for reasons unknown the agency refused to help. The health department again asked for help on Sunday morning, and again the request was denied. By that time the temperature inversion was nearly at an end. The federal government was clearly leaving Donora to its own devices. Whether help from the USPHS would have saved lives or just added to the confusion can never be known.

The USPHS finally responded to the crisis when Pennsylvania's governor, James H. Duff, and the state's industrial hygiene director, Dr. Joseph Shilen, formally requested help. The USPHS team finally arrived Thursday, November 4, more than a week after the smog had begun.

25

GAME DAY

IT WAS THREE O'CLOCK, AND STAN SAWA, KEN BARBAO, DON PUGLISI, and the other Donora Dragons players were ready for their highly anticipated matchup against the Monongahela Wildcats. The game was set to start at 3:15 on Legion Field, behind Donora High School, at the top of the hill at Fourth and Waddell. The Wildcats were underdogs, according to every sports journalist in the valley. The last time the Wildcats had beaten the Dragons on their home turf was six years before. In that game Wildcats' left end Dom Mancini nabbed a blocked pass in the final minutes of the game and ran it twenty yards into the end zone, his touchdown clutching a Monongahela victory, 12–6. Since that game, though, the Wildcats had lost every other contest to the Dragons.

Donora teacher Alice Capone was in charge of the concession stand at the game, and because of the fog she wasn't sure how big the crowd would be. "We had fifty pounds of hot dogs," Capone said. "I was worried about the hot dogs, the pop, buns, and candy [if no one showed up]."

She needn't have concerned herself; the stands were filled with an estimated seven thousand fans. How much the fans could actually see of the game, though, varied from person to person. Charles Stacey, local historian and retired superintendent of the Ringgold School District, recalled that the air that day was "relatively clear" and that "the atmosphere didn't in any way effect the players in the game." Rose Marie Iiams, a retired Donora pharmacist who also attended, agreed, saying, "I mean, it [the fog] was bad, but I think the media made it worse."

Ken Barbao played for the Dragons in that game and wore the number thirty-six on his jersey. Barbao was a five-foot-ten, 180-pound junior, just the right size, the coach told him, for a guard. He started the game at left guard and played through to the end. The fog didn't bother him at all. "It was just part of what was there," Barbao said, "and we just went right through it." The only difference between that fog and the other fogs he had experienced was that this one was "a little thicker and a little bit better spread."

The slightly older John Lignelli had a different experience. Lignelli had graduated from Monongahela High School a few years before and would become one of Donora's mayors, serving for twenty years before his retirement in 2014. Lignelli recalled, "You couldn't identify the ballplayers. You could see movement on the field, but you didn't know who had the ball and what was going on. But we stayed and watched."

Regardless of what attendees remembered of the field's visibility that afternoon, fans standing or sitting closer to the field unquestionably saw the action more clearly than did those farther away. The sportswriters in attendance apparently followed the game without difficulty. For example, the *Daily Republican*'s sportswriter, Allen Kline, described the Monongahela touchdowns this way:

Paul Anders pitched twelve yards to Bert Togni for Monongahela's first score in the initial stanza. Anders bucked over from the one in the third period and tossed 41 yards to Aaron Davis for the third score three minutes later. The final Wildcat score came on a half-yard drive by Anders early in the fourth period. [Punter] Ron Necciai converted on his first three tries, but his fourth attempt was blocked.

The sportswriters also universally failed to mention fog in their write-ups, probably because they had covered games in Donora before and were at least somewhat accustomed to the town being shrouded in smoke. Monongahela went on to win the game 27–7, an unusual loss for the tough Donora team. It was out of that game, though, that grew one of the smog's most persistent myths. According to the oft-told story, Stanley Sawa, Donora's left end, was called off the field by a loudspeaker announcement and told to return home immediately. Sawa left the game, ran down the hill to his home at 601 Fifth, rushed in, and asked the neighbor who greeted him, "What's going on? Why did you make me leave the game?"

"It's your dad," the neighbor replied. "He's in there with the doctor. It doesn't look good." Stanley was too late; his father was dead.

Bill Davidson, who had presciently written the story on the Meuse Valley fog, followed up a year later with another *Collier's* piece that covered the Donora smog. In it Davidson confirmed that announcements were indeed made. "During the game several spectators collapsed and were carried away," he wrote, "but the cases were too scattered to attract much attention. The public-address system kept announcing the names of persons who were wanted at home 'because of an emergency.'"

Stanley Sawa, however, was not one of them. Sawa played nearly the entire game and actually scored Donora's lone touchdown with just moments left, when he nabbed a short pass from second-string quarterback Don Puglisi.

No other player fits the Sawa story either. The only person to have died while the game was being played was a Donora barber named Taylor Circle, who perished at four o'clock from myocarditis, or inflammation of the heart. He also suffered from hardening of the arteries and asthma— diseases recognized now, but certainly not then, as being caused by air pollution. Circle had a son, but he didn't play on the 1948 team. He had been honorably discharged from the service in 1943, his high school football days, if any, well behind him.

The Sawa myth probably arose when stories about an earlier event began to mix with stories about the 1948 game, a contest that became known as the Smog Bowl. Sawa's father, Joseph, had suffered a stroke in 1945, three years before. As it happened, Stanley had been playing football when Joseph became ill. "I got a call in the middle of a game," Stanley later told his son, James. "I had to go home, because my dad had a stroke. I'm running up these streets in my cleats, in my football uniform, to get home." Joseph Sawa lived two more years before a heart attack felled him, shortly after noon on October 11, 1947, a year before the Smog Bowl. As stories about the Smog Bowl were told over the years, the announcements at the football game most likely became conflated with the story of Stanley being pulled from the 1945 game. That story eventually became credible and has been repeated ever since.

Not everyone in Donora was thinking about football that afternoon. Liberata Di Sanza was focused solely on her husband. Around eleven o'clock that morning, Bernardo seemed to feel better. He got out of bed and asked to talk to a priest. Having not eaten for two days, he also asked for something to eat. Hospice professionals term that kind of end-of-life rally "terminal lucidity." According to psychologist and author Marilyn A. Mendoza, "The term itself refers to someone who is dying and has been uncommunicative and unresponsive, but prior to death, becomes alert, lucid, and verbal. Family members often view this as a miracle, and as a result, often expect that their loved one will make a full recovery, only to discover that death is not far behind."

Liberata probably thought much the same, but when Bernardo's breathing worsened and he returned to bed two hours later, she became frantic. She kept calling physicians, but they were all busy. No one could come to the house, and she couldn't bring Bernardo anywhere. She felt Bernardo's hands; they were cold and damp. She could feel her beloved slipping away, and there was nothing she could do. She was with him until the end, and when Bernado died at 3:30 that afternoon, Liberata found herself alone.

Saturday was turning out to be a busy day for Donora undertakers. From 6:45 that morning until 10:00 p.m., a period of just over nine hours, eight people died.

The first was Ignace (sometimes spelled Egnace) Hollowiti. Hollowiti immigrated from Austria around 1908 and settled in Donora. He served as a fireman for American Steel & Wire and died at 6:30 a.m. He was sixty-four.

Perry Stevens died fifteen minutes later. Stevens had been born around 1894 in Alabama and, at the time of the smog, was rooming in a house owned by George and Lucinda Grimmett in Forward. Stevens had lived in the area six years and had been unemployed for the last three. He started feeling ill early Thursday morning, with dyspnea and a choking sensation. He felt better later that morning and walked down the hill to downtown Webster. He stayed only a short time before becoming short of breath again. Like many others in the valley's more rural areas, Perry died without receiving any kind of medical attention.

Suzanne "Susie" Gnora, another of Saturday's victims, did receive medical attention, but it wasn't enough. Born in 1886 in Závadka, Slovakia, Susie married John Gnora in 1907. The couple lived in Webster for about twenty-six years before moving to Donora in the late 1930s. They lived high on the hill in a small residential area called Donora Place Plan. John was a coal miner for Buds Hill Coal Company in 1948, and Susie was a homemaker. She arose early Friday morning feeling fine. "She get up and fix my lunch to take," said John, "and I go to mill. I got far to walk. Monessen. She was okay." When John returned home at five that afternoon, his wife wasn't the same. "Do what you please," Susie told him. "I can't make you supper. Do what you want. I no feel good."

John called for a doctor probably around nine o'clock that night, but no one could come. Doc Bill Rongaus was finally able to reach the house around midnight. "She was coughing severely," Rongaus said, "cold, clammy, her skin was wet. She couldn't breathe. I heard practically nothing in the chest. [She] just couldn't breath, gasping for air. I couldn't find anything else that I could attribute other than this heavy fog getting her."

He gave her an injection of Adrenalin and some cough syrup. The medicines eased her cough, but her dyspnea continued. Later that night it worsened, and she died at 8:30 a.m.

Daniel William Gardiner (typically misspelled in news reports as "Gardner") died two hours later. Born in the County of Lanark, Scotland, just south of Glasgow, Gardiner worked as a janitor for American Steel & Wire. His illness began like so many others had, with an attack of dyspnea and cough Thursday evening. William often reacted to Donora fogs that way and usually received shots of Adrenalin or morphine as treatment. A doctor arrived at some point, probably Friday, and gave him an injection and a dose of cough syrup. Gardiner's condition deteriorated and, like others, his body weakened over time and then finally gave up.

Pity poor Marcel Kraska. Born in 1882 in Poland he had been pensioned from American Steel & Wire as a wire drawer in 1946 for "an asthmatic condition which was aggravated by fog and smoke." Kraska and his wife, Victoria, were grandparents to sixteen and great-grandparents to another sixteen. All thirty-two of those children lost their grand- and great-grandparents in the same month. Victoria, who was eight years older than Marcel, died before the smog, on Wednesday, October 6. Marcel followed her twenty-four days later, a victim of the smog, at 11:45 a.m. on Saturday.

Fifty-two-year-old widower Milton Elmer Hall had been sickly for several years. He had worked in area coal mines as well as at the Zinc Works and was forced to retire for health reasons in 1936. Hall was with his son, Charles, in Webster when he died. Stricken that morning with shortness of breath, Hall's condition deteriorated quickly. Around 1:45 p.m. he turned to Charles and collapsed. He died in his son's arms.

Michael Dorincz was next. A long-retired coal miner from Austria, Dorincz became ill Friday evening with cough and dyspnea. He had been suffering from heart failure for a few years, and the smog finished him. He died fifteen minutes before the opening kickoff of the Smog Bowl.

Taylor Circle, the barber who died during the game, was eighty-one years old and the last of those who died during the daytime on Saturday, and the smog was not yet finished.

26

DONORA GOES
TO PRESS

THE TINY HOSPITAL FOR THE STEEL AND ZINC FACTORIES FIRST BEGAN
treating ill workers about four o'clock Friday, according to Eileen Loftus,
an American Steel & Wire nurse. "A worker staggered in," she said, "gasping. I had him lie down, and gave him oxygen. Then another man came
in, and another." Within a few hours every bed and exam table were filled.
The sound of wheezing, gasping, and pleading for more oxygen filled the
air that night, and again all day Saturday. In total those two days, Loftus
and other healthcare providers at the hospital cared for forty-five workers with cough, dyspnea, headache, sore throat, and various other smoke-
related conditions.

Mill officials were also busy, though in a different way. Although little
is known about smog-related activities at US Steel, it seems likely that
the phone lines between mill officials and US Steel leaders would have
nearly burned to ash. M. M. Neale was probably the first to notify corporate bosses in Pittsburgh. Neale at that point had been superintendent for
twenty-three years and knew what was expected of him. Sometime Saturday afternoon he checked the stacks himself, or told an underling to do so.

They apparently found no change in wind direction, and the decision was made to continue normal operations.

Neale checked one last time that evening, around seven, and found the smoke still dispersing "slightly north," as it had been that morning and afternoon. Neale might not have known about anyone dying in the smog before midafternoon on Saturday, but he was almost certainly notified of the situation by Burgess Chambon that afternoon. Chambon was on the phone from the time he arrived home from Pittsburgh until well into the evening, alerting everyone he thought could help. Chambon later described his activities that day:

I called every town around here to send supplies for the station and oxygen for the firemen. I even called Pittsburgh. Maybe I overdid it. There was stuff pouring in here for a week, but what I wanted to be was prepared for anything. The way that fog looked that day, it wasn't ever going to lift. And then the rumors started going around that now people were dying. Oh, Jesus! Then I was scared. I heard all kinds of reports. Four dead. Ten dead. Thirteen dead. I called a special meeting of the council and our board of health and the mill officials for the first thing Sunday morning. I wanted to have it right then, but I couldn't get hold of everybody, it was Saturday night. Every time I looked up from the phone, I'd hear a new rumor, usually a bigger one. I guess I heard everything but the truth. What I was really afraid of was that they might set off a panic. That's what I kept dreading. I needn't have worried, though. The way it turned out, half the town had hardly heard that there was anybody even sick until [Saturday] night, when Walter Winchell opened his big mouth on the radio.

Walter Winchell started his career as a vaudeville hoofer before his first writing job with the *Vaudeville News* in 1920. He reached nationwide fame as a gossip columnist for the *New York Daily Mirror*, where he worked from 1929 to 1963. His columns were so well received that he was offered a radio show on ABC in 1930. Winchell had an energetic, staccato delivery that lent an air of urgency even to stories about a celebrity's divorce or politician's bankruptcy. Over the years his own vocabulary emerged. He referred to weddings as "blessed events," expectant mothers were "infanticipating," and couples who broke up had a relationship that went "pfffft."

Winchell's radio shows famously started, "Good evening, Mr. and Mrs. America, from border to border and coast to coast and all the ships at sea. Let's go to press!" Winchell's success led to his dominance in arenas other than radio. Stanley Tucci's line in the 1998 made-for-television movie *Winchell* was only slightly more exaggerated than the real rumormonger's power at its height: "Mr. Mayor, my column gave you this office, and it can take it away."

While M. M. Neale and his bosses at US Steel that evening were trying to determine what actions they would take, Winchell went on the air and declared, "The small, hard-working steel town of Donora, Pa., is in mourning tonight as they recover from a catastrophe. People dropped dead from a thick, killer fog that sickened much of the town. Folks are investigating what has hit the area."

The news stunned residents who heard the broadcast, and word flashed around the valley to those who hadn't. Fear spread. Alice Uhriniak and the other switchboard operators handled a call volume that exploded after the Winchell show. Out-of-town relatives tried to call their loved ones in Donora or Webster, desperate to make sure they weren't among the victims. Many residents decided to leave the valley and seek cleaner air, but the thick, dark fog made travel by car nearly impossible.

At least one physician, and possibly only one, William Rongaus, had been telling his patients to leave the area, that the fog was extremely dangerous, and that smoke from the mills was to blame. "I told them to get out of town," Rongaus said.

Arnold Hirsch, a young lawyer in town, was one of those Rongaus helped. "My mother, who had not been well for years, just could not catch her breath," Hirsch recalled. "I called Doc Rongaus, and he said he just could not make it. He said, 'The whole town is sick. Even healthy fellas are dropping. Get your mother the hell out of town!'" Hirsch did as instructed and drove his mother and brother out of the valley, eastward toward the Allegheny Mountains. They all survived.

The Hirsches were lucky. For many residents, leaving town seemed a herculean task. Even if a family had friends or relatives who lived outside of the valley, and many families didn't, getting there was another matter entirely. By Friday night the fog had become so dense that even walking was fraught with danger. Yet it probably wasn't the idea of difficult travel that kept the majority of Donorans home. It was more likely a reluctance to believe in the severity of the situation, a denial of a reality they could see, taste, and smell themselves. No, Doc Rongaus must be exaggerating. This fog might be thicker than most, but valley residents had dealt with thick fogs before and had come through just fine. They would come through fine this time too.

Rongaus had also noticed that ill people who had walked or driven to Palmer Park seemed to feel better quickly. So, for a time, Rongaus and his brother hauled some sick residents in a horse-drawn wagon to the hilltop park. "Soon as we got them above the smog," said Rongaus, "they would get much better."

Rongaus knew almost from the start that smoke from the factories was at fault. He had been treating mill workers and their families for asthma and assorted other respiratory ailments since he first opened his practice. It was worse now than ever. "People were dying while I was treating them," he said. "I called it murder from the mill. I was mad, but where was I going to go for help?"

Other physicians might also have suggested that their patients leave the area, but it seems doubtful that any of them would have implied, in even the slightest way, that the metal plants were to blame. Rongaus, at thirty-four, was the youngest physician in town, though not by much. The next oldest, Chads O. Chalfant, a family doctor, was thirty-eight, and the next, Herbert Levin, was just forty. Perhaps Rongaus's youth and his being relatively new in town played a role in his being so outspoken. Like every other physician in town, however, Rongaus treated patients who depended on the plants for their livelihoods, which made Rongaus's livelihood dependent on the plants as well. He must have known that criticizing US Steel could have sparked a mill-driven boycott of his practice. So why did he rock that rich, powerful boat so directly?

Perhaps he couldn't do otherwise. From a young age Rongaus had felt a special bond with the less fortunate, the different, the people who couldn't fight back. At some point, probably in the early 1920s, Rongaus's mother noticed that her son couldn't open his mouth. Doctors determined that one of Rongaus's many ear infections had led to the formation of a painful abscess in his left mastoid bone, a thick bulge of bone behind and below the ear. In the days before the discovery of penicillin in 1928, surgeons treated the abscess by digging into the infected bone, an operation called mastoidectomy, and removing the infected tissue. During that surgery on Rongaus, it seems, surgeons inadvertently damaged the facial nerve.

"Facial nerve injury was a common complication of mastoidectomy," explained Stephen Mass, an otolaryngologist in Doylestown, Pennsylvania, "because you have to drill the bone near the ear canal, where the facial nerve runs. Doing that surgery in the twenties, when they were using chisels and curettes to remove the abscessed bone, it wouldn't be surprising that they hit the nerve."

The damaged nerve left Rongaus with a permanent numbness of the left jaw and a prominent facial disfigurement. When he talked or smiled the left side of his face moved only slightly, while the right side moved normally. He never let the disfigurement affect him outwardly, but he must have felt acutely aware of its presence. As he grew into his thirties and forties the disfigurement became more pronounced; the muscles of his right jaw grew significantly stronger than those of his left jaw, which lacked normal movement. The condition must have made him highly attuned to the needs of his patients. More compelling in Rongaus's defense of the underdog, though, was the relationship he had with his brother Walter. Bill idolized his brother. Actually the whole family did. "Walter was put on a pedestal by the entire family," said Bill's daughter, Nancy Rongaus Cherney, "and my dad was lockstep with his siblings and parents."

Walter was the first Rongaus child born in the United States. Bill was second. Walter was a surgeon. Bill was a family doctor. Walter was an accomplished musician, playing clarinet and saxophone in the Casa Loma Orchestra, a well-known dance band, during summer breaks from medical school. Bill played no instrument, couldn't sing, and didn't dance. Walter served as a frontline surgeon during World War II, operating "morning, noon, and night" on soldiers wounded in D-Day's Operation Overlord. Bill's hearing loss kept him in Pittsburgh, far away from the front.

"My father always felt inadequate because he couldn't serve," said Cherney. "He wanted to be a hero and a warrior like his older brother, Walter."

When smog rolled into Donora, Bill might have seen the disaster, subconsciously at least, as a chance to stand out from the crowd, to make his presence in the community truly felt, and, perhaps, for once, to outshine his perpetually shining brother. Intended or not, Bill Rongaus would emerge from the tragedy as much a hero and warrior as his beloved brother.

––––––––––––

The reverberations from Winchell's broadcast were still ricocheting around the valley when Edward Roth speculated that perhaps the smog was clearing. "I'd had a call about noon from a woman who said two men roomers in her house were in bad shape," he said. "It was nine or nine-thirty by the time I finally got around to seeing them. Only, I never saw them. The landlady yelled up to them that I was there, and they yelled right back, 'Tell him never mind. We're OK. now.' Well, that was good enough for me. I decided things must be letting up. I picked up my grip and walked home and fell into bed. I was deadbeat."

Down at the makeshift emergency center on McKean Avenue, Cora Vernon and her team had prepared to treat the sick. "We were ready for anything and prepared for the worst," said Vernon. "We even had an ambulance at our disposal. Phillip DeRienzo, the undertaker, loaned it to us. But almost nothing happened. Altogether, we brought in just eight patients. Seven, to be exact. One was dead when the car arrived."

That individual was a fifty-five-year-old housewife named Barbara Chinchar. Born in Austria on May 15, 1893, Chinchar had lived in the area for forty-two years. She had awakened Thursday morning with a head-ache, which gradually worsened despite treatment by a physician Friday and again Saturday. Around dinnertime Saturday she began feeling short of breath, perhaps thinking that the tuberculosis she had two years before was acting up again. It wasn't tuberculosis, it was the air, and she died at ten o'clock that night, apparently in the back of a hearse on the way to the community center for treatment.

Vernon didn't treat Barbara Chinchar, but she did treat the other seven patients, sending three to the hospital and the others home. After that, everything settled down. "It was really very queer," Vernon explained. "The fog was as black and nasty as ever that night, or worse, but all of a sudden the calls for a doctor just seemed to trickle out and stop. It was as though everybody was sick who was going to be sick. I don't believe we had a call after midnight. I knew then that we'd seen the worst of it."

Roger Miles Blough was a successful New Jersey attorney before joining US Steel and working his way up to general counsel. He was at US Steel's corporate office in Pittsburgh at three o'clock that morning, as the company's executives at long last decided to act. Blough called Neale and told him to dead-fire the smelter furnaces, temporarily stopping the furnace from producing zinc. Neale supervised the dead-firing to completion, and by seven o'clock Sunday morning was in his office, which at that point "resembled Grand Central Station."

Blough hadn't made the decision about the dead-firing; that probably would have been the job of the chairman, Benjamin Franklin Fairless. Why the decision hadn't been made sooner has been a matter of conjecture ever since. Blough might have supported earlier intervention, urged delay, or remained neutral throughout the discussions. Whatever position he took seems not to have mattered much to US Steel.

Seven years later, in May 1955, Blough would be named chairman and chief executive of the company. That same year, just two months later, President Dwight D. Eisenhower would sign the first comprehensive air pollution legislation in history, authorizing federal assistance for states to the tune of $16.5 million for "research, training, and technical assistance."

One more person would die during the smog, a miner from Sunnyside named John R. West. Born May 5, 1892, in Americus, Georgia, West had been a coal miner for forty years. He took ill Friday afternoon, returning from work at four o'clock with chest pain, headache, and trouble breathing. He couldn't lie flat or sit up without dyspnea, so he knelt on the floor with pillows around him; he could breathe better that way. A Monongahela physician, Norman Golomb, traveled to West's house twice and gave him two injections each time, one of Adrenalin, certainly, and perhaps one of morphine to treat the chest pain. The treatments didn't last, and West succumbed to the smog at 5:00 a.m. on Sunday, still kneeling. West's death was the last of the weekend.

Golomb, who signed West's death certificate, did something only one other physician treating a smog victim did. He noted West's cause of death as "bronchial asthma," and then wrote under the "Due to" heading,

FIG. 26.1. This map shows where each victim lived and, for all but three, died. Courtesy of Donora Historical Society.

The labels on the map read:

Forward
George Hvizdak, 52
Peter Stankovich, 67
John West, 56

Webster
Milton Hall, 52
Emma Hobbs, 55

Fellsburg
Ida Orr, 58
Thomas Short, 81

John Cunningham, 63

Ignace Hollowiti, 64

Susan Gnora, 62

William Gardiner, 66

Taylor Circle, 81

Andrew Odelga, 69

Sawka Trubalis, 65

Michael Dorincz, 84

Ivan Ceh, 69

Marcel Kraska, 65

Bernardo Di Sanza, 67

Perry Stevens, 54

Barbara Chinchar, 55

Jane Kirkwood, 76

"Abnormal weather conditions." He was the first to ascribe a death specifically to the smog.

Still the smog wasn't finished.

———————————

Blough and the US Steel executives, along with Neale and L. J. Westhaver, another mill official, continued to communicate through the night. They called in chemists from the Industrial Hygiene Foundation, part of the Mellon Institute in Pittsburgh, to examine the composition of air in Donora. The first chemists arrived around 6 a.m. Sunday. Chemists from the University of Pittsburgh arrived later, as did two industrial hygienists from the state's department of health.

The teams obtained six samples of air quality near the Zinc Works and sent them to Pittsburgh for testing. The head of the Pennsylvania team, Joseph Shilen, didn't expect much from the results. "The greater part of the fog had passed by the time we got into operation," said Shilen.

The morning brought with it a harried and at times contentious meeting of town and mill officials. Burgess Chambon brought the meeting to order at around ten o'clock. The president of the town's Board of Health, along with the board's executive secretary and owner-operator of the Physicians Exchange, Elizabeth Ostrander, reported on the number of sick and dead, as well as the number of ill residents remaining in area hospitals: forty-five. Chambon told the assembled that he had asked superintendents Neale and Westhaver to bank the furnaces immediately and to keep them banked until the Pennsylvania Department of Health declared that operations could resume safely.

"They said they already had," recounted Chambon. "They had started banking the fires at six that morning. They went on to say, though, that they were sure the factories had nothing to do with the trouble. We [town officials] didn't know what to think."

Others at the meeting were sure that they knew and blamed the Zinc Works. Arguably the most outspoken of them was Doc Bill Rongaus, who called the smog's effects on residents "just plain murder." He noted that air pollution in Monessen, Homestead, and other neighboring mill towns hadn't caused deaths of the magnitude seen in Donora. To bolster the point he also remarked, erroneously as it turned out, that the deaths had occurred within blocks of the Zinc Works. (Only one person, John Cunningham, lived reasonably near the Zinc Works.)

Superintendent Neale consistently defended his mill from attack. Neale pointed to the long history of the smelter and said that no deaths from its effluents had been identified in all of that time. He said that US Steel had banked its furnaces as a "gesture of concern" for the community and "not an admission of responsibility."

Neale the human was almost certainly concerned about his neighbors; he had been living with them for many years and working diligently for them on various committees. Neale the businessman, however, made sure that neither he nor any of his subordinates even intimated that his company was at fault. The event, he said numerous times, was an "act of God," a weather event that no human could have prevented. In addition, Neale made clear that many people in the valley used coal stoves for heating, as did several industries. With so much smoke being emitted into the air, how could anyone conclude which specific sources of that smoke were to blame? No, no, it just wasn't that simple. And the mills were most certainly not to blame.

Chambon, the town's leader, the man who almost lost his own mother to the smog, stood solidly behind Neale. He tried to deflect any blame being placed on the factories, saying that placing blame was irrelevant and that this was a time for everyone to pull together. "These deaths were a blessing in disguise," he said. "They brought this matter to a head. Let's all pull together and eliminate this trouble."

He admitted, however, that the whole situation had taken a toll on him and that the meeting that morning "was the worst. It wasn't just that the fog was still hanging on. We'd begun to get some true facts. We didn't have any real idea how many people were sick; that didn't come out for months. We thought a few hundred. But we did have the number of deaths. It took the heart out of you. The rumors hadn't come close to it. It was eighteen."

The town and factories could have pulled together immediately, but it wouldn't have saved Sawka Trubalis. Sixty-six years old, Trubalis had immigrated from Russia in 1904 and went straight to Donora. He and his wife, Pauline, lived in a modest home at 438 Sixth Street. Sawka worked in

American Steel & Wire's open hearth department until late 1947, when he retired, most likely due to having developed myocarditis over the years. In patients with chronic myocarditis the heart struggles to beat hard enough to pump adequate amounts of blood throughout the body. Any stress on the heart, such as that from a suffocating smog, can overwhelm the heart completely.

Trubalis must have been a rugged individual, and perhaps stubborn as well. He first fell ill Wednesday evening with dyspnea. The episode didn't last long, and by the next morning he felt much better. The dyspnea returned Friday morning around 7:30. His chest hurt, he found that he couldn't breathe when lying flat, and he now had a cough as well.

Levin, the physician who had treated Mother Chambon, also treated Trubalis sometime that day. When Levin arrived he found Trubalis cyanotic and in shock. His breathing was rapid and shallow. Levin couldn't hear the man's breath sounds; there wasn't enough air flowing into the deeper parts of the lungs. Levin tried to obtain a blood pressure, but he couldn't hear anything through his stethoscope. Trubalis was a sick, sick man.

Levin gave him an injection, or perhaps just a pill. Whatever it was, it helped for a time. Trubalis's condition deteriorated again on Saturday. An ambulance came at midnight Saturday and brought him to Charleroi-Monessen Hospital. Nurses there immediately placed him in an oxygen tent. He was almost certainly treated with Adrenalin and theophylline, but no treatment at that point could have saved him. Trubalis died Sunday at 12:30 p.m., having survived the impenetrable fog for an incredible four and a half days.

And the smog continued to kill.

27

BLESS THE
RAINS

THE STAFF AT CHARLEROI-MONESSEN HOSPITAL THAT WEEKEND TREATED
patient after patient struggling to breathe, coughing, and complaining
of chest pain, headache, and nausea. The hospital's few emergency beds
filled up quickly, and so too did the waiting room chairs. Patients were
treated and either released or admitted, and as soon as one patient left
a bed, another one took their place. The nursing staff rushed to transfer
patients out of the emergency room and into a waiting bed on the wards.
The empty ward beds soon filled up as well, and physicians began discharging less ill patients so that more gravely ill patients in the emergency
room could be admitted.

Even if the hospital had enough empty beds to accommodate all the
sick from Donora and Webster, it didn't have anywhere near enough
oxygen tents for everyone who needed them. Nursing supervisors called
other hospitals in the area, even some in Pittsburgh, to obtain more tents.
The hospital almost certainly would have exhausted their in-house supply
of Adrenalin; no hospital its size would have kept enough on hand to cope

with such an emergency. So calls would have been made to pharmacies in Charleroi and surrounding areas.

The chaos of that little hospital must have been overwhelming for patients and family members who entered through the doors. Nurses in brilliant white caps and crisply starched white uniforms hustled up and down hallways, into and out of rooms, and back and forth behind the main desks. Doctors in white lab coats huddled by bedsides, bent over lab reports, and gazed up at X-rays hung on a light board. Rooms and hallways filled with the sound of dry, hacking coughs and the high-pitched whistling of wheezes, a cantata of imperiled voices muffled only by oxygen masks and plastic oxygen tents. A mixture of odors struck everyone coming into the hospital, alcohol and acetone, ash and soot, cigars and cigarettes, and a host of other odors too faint or unfamiliar to identify. Housekeepers, technicians, maintenance men, secretaries, ambulance drivers, and family members with tense, worried faces only added to the mayhem inside the hospital walls.

Eighty-one-year-old Thomas Amos Short entered into that mayhem early Sunday morning, October 31. A retired coal miner, Short was born in England's Gloucestershire County. He immigrated to the United States in 1883, at age sixteen, and arrived in Donora five years later to mine coal. Short had awakened with dyspnea and chest tightness at 4:00 a.m. on Friday. He apparently received some type of medical care later that day, which seemed to help.

At around five o'clock Sunday morning the dyspnea returned, worse than ever. He became disoriented, even delirious. He was brought to Charleroi-Monessen and immediately placed into an oxygen tent. He was cyanotic, severely dyspneic, and had a temperature of nearly 104 degrees. Doctors were able to reduce the fever, but Short continued to be critically short of breath. He didn't respond to any treatment and died eight days later, at 5:05 a.m. on November 8, still inside his oxygen tent.

Chads O. Chalfant, one of Donora's eight family physicians, signed the death certificate. Chalfant noted that Short died of "asthmatic bronchitis," a commonly noted cause of death for smog victims. Then, in the "Due to" section, Chalfant wrote, "(Smog)." Like Golomb's note on John West's death certificate, Chalfant had also come strikingly close to outwardly blaming the metal plants for someone's death. Although Chalfant actually put a name to the beast and even capitalized it, he surrounded the word with a set of parentheses, like a detective who announces the murderer's name and then whispers, "maybe."

One victim has been largely overlooked since the smog: George Hvizdak (sometimes spelled Weisdack). Hvizdak was a fifty-two-year-old immigrant from Úbrež, Slovakia, who in 1948 was retired from the zinc smelter and farming his land in Webster. He began feeling poorly Saturday morning, October 30, with dyspnea and chest pain. With a history of asthma, he might have assumed that he was having a typical asthma attack, but his condition continued to worsen. He was hospitalized on November 13, nearly two weeks after the smog ended. He responded well to treatment and was released five days later.

His recovery didn't last, though, and he was readmitted December 3 with the same symptoms. This time Hvizdak continued to fail and developed a lung abscess while in the hospital. Physicians gave him penicillin and placed him in an oxygen tent, but his body proved too fragile from being so sick for so long. He surrendered to the smog at 10:45 p.m. on December 22.

The smog had killed its last victim.

August Chambon called another meeting with the same people for two o'clock Sunday afternoon, October 31. Rongaus again railed against the plants. "In my opinion," he told the assembled, "something should be done about smoke coming from the mills. The fumes are killers. They are silent killers." Others probably thought the same but were too reluctant to speak out to reporters.

Chambon said about the meeting, "It was the same thing all over again. We talked, and we wondered, and we worried. We couldn't think of anything to do that hadn't already been done. I think we heard about the nineteenth death [probably Sawka Trubalis] before we broke up." He had no answers for the questions swirling in his head. *What would happen if the fog didn't lift? What could the council do? What could anyone do?* Yes, the community center had been turned into a first aid center, and yes, doctors, nurses, firefighters, and ambulances had arrived from neighboring towns and had been helping to treat the sick, but beyond that, what more could anyone do if the fog didn't lift?

Lift the fog finally did. A mass of warm air from the west began to flow into the valley early Sunday morning. Reports on when rain arrived vary, but a drizzle began to fall probably in the early afternoon, not earlier, as some reports suggest. The arrival of wind and rain led to a breakup in the temperature inversion. The fog began to dissipate. By the time Chambon called the afternoon meeting to an end and walked outside, he noticed a welcome change. "When we came out of the building," Chambon said, "it was raining. Maybe it was only drizzling then—I guess the real rain didn't set in until evening—but, even so, there was a hell of a difference. The air was different. It didn't get you any more. You could breathe."

———

A few smoke control officials from Pittsburgh arrived in Donora that afternoon and supported Rongaus's repeated claims about the cause of the deaths. The officials explained that sulfur monoxide, sulfur dioxide, and sulfuric acid emitted from the Zinc Works smokestacks built up in the valley to toxic levels, just as had happened in the Meuse Valley in 1930. Dr. Isaac Hope Alexander, director of the Pittsburgh Department of Health, gave his summation as well. "The smoke from the Zinc Works at Donora," he proclaimed, "is of a very toxic variety."

From that point forward Doc Bill Rongaus was no longer alone in his condemnation of the metal factories.

———

Sometime in the early morning hours of Sunday, Bill Schempp stopped bumping his way up and down foggy streets. He returned home to Gladys and a soft, comfortable bed. Calls for Roth, Koehler, Chalfant, and the other physicians diminished over Saturday night and finally ceased. Each man went his separate way to rest and recover from an agonizing weekend.

Superintendents Neale and Westhaver might have rested as well, but it wouldn't have been for long. Scientists from the US Bureau of Mines arrived in town Monday, and officials of the steelworkers union met and demanded an investigation into the smog and the deaths it had caused. There would be multiple investigations and many lawsuits filed, but residents of the valley weren't concerned with any of them that day, nor, especially, the next.

The afternoon drizzle turned that evening into a steady rain, ridding the air of whatever toxins remained and washing them into the muddy Monongahela. The rain lightened on Monday and stopped around noon. Residents throughout the valley started their Monday as they typically did, as if the prolonged smog they had just survived was perfectly normal. When the rain stopped, "every housewife who wasn't sick in bed rushed to get her washing done that day," said resident Helen Liwak, "while the air was clean."

Charleroi-Monessen Hospital began discharging smog patients, the first of whom was Haymen "Hymie" Kohn, a sixty-four-year-old blacksmith at the wire mill. Alvie Woodburn and James Spence, both of Donora, were also released on Monday, with at least seven more residents still being treated. Hospitals in Monongahela and Pittsburgh likewise began discharging patients sickened by the smog.

Later, long after the last victim, George Hvizdak, was laid to rest in Saint Michaels Cemetery in Donora, investigators would determine that of the approximately 13,000 residents of Donora and 1,000 residents of Webster, roughly 6,000 had been sickened by the smog. Hundreds had been hospitalized or left town to escape the smog. And an untold number of townspeople survived due to the courageous, dedicated, inexhaustible work of the town's doctors, nurses, firefighters, and ambulance drivers.

Tuesday morning, November 2, simply sparkled. A cerulean sky commingled with puffy white clouds. Residents emerged from their home inhaling crisp, clean air—clean for Donora, at least. It seemed a perfect day, spoiled only by a series of funerals and burial processions for eight of the smog victims.

Rudolph Schwerha was busy that day. He managed the wakes and funerals of eleven of the smog's twenty-one victims. He surely felt the irony of burying victims of a dark, toxic fog on such a bright, sunny, glorious day. "It was like a day in spring," Schwerha said. "I think I have never seen such a beautiful blue sky or such a shining sun or such pretty white clouds. Even the trees in the cemetery seemed to have color. I kept looking up all day."

Four smog victims were buried in Saint Dominic's Cemetery, perched high on a hill above Donora. Nearly the entire cemetery is built on a slope. Gravestones are arranged in neat rows, like seats in an amphitheater, each afforded a view of the valley below. At some point that morning a procession of mourners made their way along the hillside to the grave of Peter Starcovich, the third victim of the smog. Men in dark suits and women in woolen overcoats passed graves marked by tall white crosses or stout granite tombstones. The mourners walked softly over grass already tamped down by gravediggers.

As a Saint Dominic's priest prayed over Starcovich's grave and mourners bowed their heads in respect, puffs of light gray smoke from smelter smokestacks filled the valley in the distance. For anyone who cared to look, the smoke proved a stark reminder that the furnaces were not gone. They were lying in wait. Soon they would be recharged and brought back to full power.

When that happened the surviving citizens of this industrious little section of the Monongahela Valley would once again live, work, and breathe in the soot-filled, ash-laden, income-generating air they knew so well. Life would return to normal. Journalists and photographers who had arrived in town to report on the tragedy would by then be gone, and the nation would forget about Donora.

But the nation wasn't ready to forget.

———————

Fog-free air in the valley brought with it an assortment of "experts" who descended on Donora and gave their frank, considered, and ultimately wrongheaded opinions on the causes of the smog. One such expert, Duncan A. Holaday, determined almost immediately that "there was no evidence found that will incriminate any one particular plant as being the source of the atmospheric contaminant over the weekend." He gave the pronouncement at a news conference on November 4, not even a week after the first smog victim, Ivan Ceh, had died. Holaday further announced that "there will be no hazard to the people of this community to resume operations [of the mills]."

Holaday might be forgiven for his erroneous pronouncements. He possessed no medical background, though he held degrees in chemical engineering and biochemistry. He also had little or no experience in air pollution of the kind found in Donora. His career to that point, and which continued to his retirement, focused solely on radiation and its effects on the health of uranium miners. He was, in fact, a leader in that field.

Forgiving Jospeh Shilen, head of the industrial hygiene division of Pennsylvania's health department, might be more difficult. Shilen was a physician of some repute in the area. He had served as director of the Tuberculosis Hospital in Pittsburgh before joining the state's health department in 1942. In his role as director he oversaw numerous health surveys, announced a variety of initiatives including one urging the industrial community to place more physicians in factories critical to the war effort, and issued many warnings on dangerous chemicals in commonly used products.

The incident in Donora, however, was far more complex than any of those efforts. It involved myriad deadly chemicals, enormously complicated metal-making operations, and the many subtle, sometimes intractable facets of small-town politics. For whatever reasons, Shilen supported Holaday and agreed that the Zinc Works and steel mills could safely restart. Superintendent Neale accepted their authorization and ordered his mill to return to full production.

At that point the town seemed split into two main factions. One followed the lead of mill officials, who claimed that the smog had been an act of God, that the factories had been operating without incident for more than thirty years, and that although the deaths were tragic, the factories couldn't and shouldn't be held responsible. This faction believed that the people who died were old and sickly, and if they weren't old, well, they must have had something wrong with them. Effluents from the factories were absolutely not at fault.

"The Zinc Works didn't have anything to do with [the deaths]," said Ken Barbao, a lifelong Donora resident. "Not *our* mills. Maybe the Monessen mills." Monessen was home to the Monessen Works, a steel mill that operated three blast furnaces. "I can remember walking down the street [Saturday] night," recalled Barbao, "and I could smell that Monessen mill." He was undoubtedly recalling another evening; there was virtually no wind that night, and the breezes that appeared on Sunday came from the west, not the south, where Monessen is located. Regardless, he believed throughout his life that the Donora plants had nothing to do with the deaths.

The Catholic bishop of Pittsburgh, Austin Purdue, believed the same. He and his wife visited a local vicar named George Davidson on November 2, 1948, two days after the smog had cleared. Purdue wrote a letter of support later that day to Zinc Works Superintendent M. M. Neale, saying, "George reported that chemical analyses have showed that the deaths had nothing to do with local industry. George stated that it was simply a vacuum or lack of air, which could happen in any of our river towns. He was very emphatic in his information." Whether Neale, an influential member of Davidson's parish, was the source of that emphasis can never be known with certainty.

The other faction in the area consisted of those who agreed with Rongaus, that the mills were responsible for the deaths, that smoke from the mills had been making people sick for years, and that the zinc smelter was by far the worst offender. They believed that the deleterious effects of the smoke could be denied no longer. Angelo F. Natali, a resident of Webster, immediately wrote a letter to Rongaus on November 1 to express his gratitude and support, saying, "I want to thank you for what you are trying to do for the people of Donora and Webster, regarding to the *fumes* in the air. Do not back down on your statement which is a *fact*. I will help you in any way I can. May God Bless you and give you faith to carry on."

Donora pharmacist Rose Marie Iiams could see both sides of the debate but came down firmly, if subconsciously, on the side of the mills. "The people that died were sickly," she said. "That was really the problem. My mother lived to eighty-six, and I lived through it. My dad lived through it. I walked to work every day. I walked to work, and I walked home at noon, and I walked back in the afternoon, and back home at five o'clock." *So it wasn't the smoke that had killed people*, she thought. *They died because they were too weak.*

Most of the people who agreed with Rongaus, though, wouldn't speak out about it. "A lot of them said they were not sick afterwards," Rongaus said later, "and I know they were goddamned liars, because I treated them. They were afraid of losing their jobs."

David Lonich, a local historian and former teacher in the Ringgold School District, offered a different reason. "My father lived through the depression, invaded Normandy, liberated Buchenwald, and was nearly killed in the Zinc Works," he said. "These were tough people, and as ironic as it sounds, their toughness led, in part, to them downplaying the dangers of the polluted environment they worked in. They were realistic about what they could accomplish against US Steel and were acutely aware of the consequences of their actions."

The residents didn't need to say anything. The federal government would soon become involved, and from that point on matters would turn from being a local story to a national one.

THE BLAMING
GAME

INVESTIGATIONS BEGAN ALMOST IMMEDIATELY, THE MOST PROMINENT and comprehensive of which was an in-depth survey by the US Public Health Service. The agency assigned the study to James G. Townsend, a physician and director of the Public Health Service's Division of Industrial Hygiene. Townsend sent a team to Donora to study the effects of smog on health, a subject that had not received the level of scrutiny it surely deserved. Leonard Scheele, US surgeon general at the time, said that the team's final report showed "with great clarity how little fundamental knowledge exists regarding the possible effects of atmospheric pollution on health."

A survey team of twenty-five specialists arrived on Monday, November 30, 1948, and opened a home base in the community center. The team included nurses, physicians, scientists, chemists, statisticians, dental surgeons, engineers, meteorologists from the National Weather Service, and even a veterinarian, Arthur H. Wolff, to study the effects of smog on animals. The team spent five months in the valley. Nurses interviewed

FIG. 28.1. Paul P. Hageman pets his dog Sputzy while being interviewed by a
public health nurse after the smog. Hageman was one of thousands sickened by the
smog and received an injection from one of the doctors that weekend. Hageman
died in 1968 at age eighty-eight. Courtesy of Donora Historical Society.

residents of Donora and Webster. Physicians examined many of them,
tested their blood, and in a few cases, had X-rays taken of their lungs. They
reviewed medical reports of all deceased residents, as well as thirty-two
of the fifty patients who had been hospitalized during the smog. The team
also performed autopsies on four victims: Bernardo Di Sanza, Michael
Dorincz, George Hvizdak, and Andrew Odelga.

Chemists, pollution scientists, and other experts took air and soil samples to identify the concentration of a wide variety of toxins. Engineers
examined mill processes from beginning to end and tested air samples
from a variety of locations within the factory complex. The team tried to
replicate the conditions of the last week of October but were unable to do
so. Even without precise replication the team found evidence of extremely

toxic elements in sufficient concentrations to affect human health. The foremost offender spewing those toxins into the air was, as residents had predicted, Donora Zinc Works.

Over the six days of the smog, the team estimated that the Zinc Works alone threw a staggering 192,000 pounds of particulate matter, 84,000 pounds of zinc, 37,000 pounds of nitrogen oxides, 309,000 pounds of sulfur dioxide, and more than 385,000 pounds of carbon dioxide into the air. Altogether the residents were exposed to more than a million pounds of toxic matter.

Fluorine was considered, for a time—and even now, by a few—the main culprit. Philip Sadtler was a strong proponent of the fluorine theory. Sadtler had graduated the year before the smog with a bachelor's degree in chemistry from Lehigh University in Bethlehem, Pennsylvania. His father and, especially, his grandfather had been leaders in the field of chemical engineering, his grandfather serving as the first president of the American Institute of Chemical Engineers.

The younger Sadtler provided his opinion on the smog in a letter to Rongaus's brother Walter just four days after the smog ended. The letter quoted two paragraphs, picked up in their entirety from an industrial hygiene journal, that described symptoms of people who had been poisoned by fluorine. After those paragraphs, the newly degreed Sadtler wrote, "There is no doubt in my mind at this time that the trouble was caused by fluorine."

Sadtler's pedigree might have been excellent, if limited in depth, but his inexperience showed. His letter listed symptoms of fluorine poisoning never or rarely noted among Donora's smog victims. All the remaining symptoms are commonly found in patients with asthma, underlying cardiac disorders, and assorted other pulmonary conditions, not just in fluorine toxicity. In any case the fluorine theory never gained traction and, in fact, was frequently criticized. The public health team measured a total of just 108 pounds of fluorine being released by the Zinc Works over six days.

The Zinc Works might have been the main polluter, but others also contributed to the deaths and illnesses that weekend. The report noted that Donora's blast and open hearth furnaces were "major contributors."

Coal stoves being used to heat homes and businesses and the number of trains running throughout the mill complex also added significantly to valley pollution.

The report noted that nineteen deaths had occurred in October 1945, a number higher than in any other month of the preceding four years. At least fifteen of those deaths stemmed from disorders of the heart or lungs, suggesting that valley residents had experienced severe smog that year as well.

Finally, the report made ten overall recommendations, nine of which focused on reducing emissions from various zinc and steel factory operations. The tenth recommendation reveals that the study team gave at least some thought to what might happen in the future. The recommendation read, "Establish a program of weather forecasts to alert the community of impending adverse weather conditions so that adequate measures can be taken to protect the populace."

Buried in a sub-report called "Meteorological Conditions and Atmospheric Contaminants" was this recommendation: "The industries in the Donora area and adjacent communities should curtail production during a stable-stagnant valley air condition." Why a stronger recommendation was never made, such as to immediately bank the furnaces at the first sign of a temperature inversion, or at least consider it, remains unexplained.

An official of American Steel & Wire, Clifford F. Hood, agreed with the report and said that "a great deal of work and research remains to be done in the future to study the causes of air pollution in the industrial communities of America and to bring about the necessary remedies." Hood, who would be named president of US Steel in 1953, also maintained that American Steel & Wire had "cooperated fully with all properly accredited agencies in an effort to develop more information on the entire subject. It is our intention to continue this policy of cooperation. It is our desire to remain good neighbors in Donora and the surrounding area."

James Townsend lauded the report for its clear victories. He said, "The Donora study has erased any doubts that contaminants in the air can exact tolls on the health and lives of people."

The report never explicitly blamed US Steel for its role in the tragedy; Townsend was too circumspect for that. He claimed in the report's conclusion that "it seems reasonable to state . . . that while no single substance was responsible for the October 1948 episode, the syndrome [illnesses and deaths] could have been produced by a combination, or summation of the action, of two or more of the contaminants."

That vague, indecisive phrasing—"seems reasonable" and "could have"—proved effective in protecting the plants but did little to protect the residents. Townsend was much more direct in his call for more research into air pollution, saying, "The lack of fundamental data on the physiologic effects of a mixture of gases and particulate matter over a period of time is a severe handicap in evaluating the effects of atmospheric pollutants on persons of all ages and in various stages of health."

The report made no recommendations suggesting that the height of the smokestacks be increased to a level above the hilltops on either side. Even though the Townsend team "threw every resource possible into an investigation of the Donora smog," its report abstained from making the kinds of recommendations—apart from a few nonspecific suggestions for Donora plants alone—that might have spurred metal-mill owners elsewhere to implement anti-pollution measures in their plants and smelters.

By avoiding a direct conflict with US Steel, one of the nation's largest and most powerful companies, the report essentially gave license for metal makers to continue business as usual. The report was a toothless, clawless jackal trying to scare away a hungry leopard.

————

Numerous smaller studies were conducted after the smog, among them a creative, industry-disrupting study whose primary author was toxicologist Mary Ochsenhirt Amdur. Amdur was born in Pittsburgh on February 18, 1921, and grew up in Mount Lebanon, a community about eight miles southwest of Steel City. With a chemistry degree from the University of Pittsburgh and a doctorate in biochemistry from Cornell, she found herself in 1949 working for the renowned Philip Drinker at Harvard University's School of Public Health.

Drinker, along with a physician colleague, Louis Agassiz Shaw, had invented the iron lung for victims of polio in 1928. The invention and the number of children saved by it conferred on Drinker international status. After the Donora smog, Drinker's lab received funding from American Smelting and Refining Company to look into the event. The company owned numerous mines and smelters in the West and wanted Drinker and his team to demonstrate that chemicals emitted by the company's smelters were essentially harmless, and so had little to do with what happened in Donora.

Drinker had investigated the Meuse smog tragedy in the 1930s and believed, wrongly, that a similar event couldn't possibly happen in the United States. He wrote in 1939, "Our stacks emit the same gases as did Belgium's. But fortunately, so meteorologists tell us, we have no districts in which there is even a reasonable chance of such a catastrophe taking place."

Drinker had been an industry apologist for years, and in fact had criticized comments made by Clarence A. Mills a year after Donora. Mills, a pioneering environmentalist and professor of experimental medicine at the University of Cincinnati, had assailed some of the results of the Public Health Service study, writing in the journal *Science*, "They spent months analysing the valley air for poisons, but failed to calculate the concentrations probably present during the killing smog a year ago, when an inversion blanket clamped a lid down over the valley's unfortunate people."

Although Drinker agreed that the public health study was flawed, he lashed out at several of Mills's comments, at times sounding like a petty child. "Dr. Mills grumbles a bit because a grant he requested from the PHS for a pollution study was refused." Drinker also claimed, without evidence, that zinc factories similar to those in Donora "have been operating all over the world for years, and the health record of men inside the plants and of people in the communities has not been adversely affected."

Perhaps Drinker should have been conducting the American Smelting studies himself, but instead he gave Amdur the assignment. He told her to investigate sulfuric acid's effects on human lungs. Drinker assumed she would return with results satisfactory to the smelter industry. He was wrong again.

Amdur was a superb scientist, and an utterly conscientious one. She conducted several unique experiments in her backyard on the effects of inhaled sulfuric acid on guinea pigs, which, like humans, breathe primarily through their mouths. Amdur found that sulfuric acid caused significant and potentially fatal lung damage in the animals. In December 1953 she presented a paper on her experiments to a meeting of the American Association for the Advancement of Science. Her presentation caused no controversy.

When officials at American Smelting caught wind of the presentation, however, they were, to say the least, displeased. Amdur was scheduled the following June to present the same paper at the American Industrial Hygiene Association conference in Chicago. At that meeting, however, two leather-jacketed goons slipped into an elevator with the scientist. Stepping close to her, one thug said, "Hey, Mary, where you going? You're not going to deliver that paper, are you?"

Amdur did indeed deliver that paper, much to the consternation of American Smelting executives, who immediately pressured Drinker to squelch Amdur's findings. The day Amdur returned to Harvard from Chicago Drinker told her to withdraw her paper, which had already been accepted for publication by the prestigious British medical journal, *The Lancet*. She refused, and Drinker fired her. The paper was never published.

Amdur would go on to a stellar career as a pioneer in the health effects of air pollution and is commonly referred to today as the Mother of Air Pollution Toxicology. Her research in the fifties and sixties, as well as the Townsend report in the immediate aftermath of Donora, served to inform activists, physicians, researchers, and lawmakers on the effects of air pollution on human health. It would take many years to develop clear and effective air pollution control legislation, and the man who would start that critical process had, in 1948, just been elected president.

Harry Truman defeated Thomas E. Dewey the same day that smog victim Peter Starcovich was buried. The Donora disaster had landed on Truman's plate of issues before he ever sat down to breakfast. He soon authorized a first-of-its-kind meeting, the United States Technical Conference on Air Pollution, held May 3, 1950, at the Wardman-Park Hotel in Washington, DC.

Truman headlined the conference and extolled the attendees on the dangers of air pollution and the urgent need to learn more about how pollutants affect the body:

With the increasing industrialization of the United States, contamination of the air around us has become a serious problem, affecting all segments of our population. Air contaminants exact a heavy toil. They destroy growing crops, damage valuable property, and blight our cities and the countryside. In exceptional circumstances, such as those at Donora, Pennsylvania in 1948, they even shorten human life. We need to find out all we can about the relationship between air contaminants and illness.

From that historic conference was born the first federal legislation addressing air pollution: the Air Pollution Control Act of 1955, signed into law by President Dwight D. Eisenhower. The act provided $5 million for research into the health effects of air pollution and, more important, signaled that the federal government was now invested in air pollution and its causes. The act underwent several changes over the years, each time being strengthened from simply investigating air pollution to setting and enforcing air quality standards.

The act known today as the Clean Air Act was signed into law by President Richard M. Nixon on the last day of the year in 1970. That law dramatically revised the 1955 act and set strict limits on air pollution from existing entities, regulated emissions from new entities, funded research into noise pollution in large cities, and opened the door for American citizens to bring lawsuits against any entity, including the government itself, that they believed had violated the law.

The year 1970 also saw the creation of the Environmental Protection Agency. Nixon named his young assistant attorney general, William D. Ruckelshaus, as EPA director. Ruckelshaus was just thirty-eight years old and would go on to have a stellar career in government, including two stints as EPA director and one as FBI director. (He was famously fired in 1973's Saturday Night Massacre, one of several damaging scandals during Nixon's final years in office.) Ruckelshaus's reputation for honesty and competence lent the EPA an authority it needed to exert control over the nation's largest industries, with some of the most powerful lobbies in history, limiting how much they could pollute the country's air, land, and water.

Without little Donora and the tragedy that occurred there, who knows how soon the government would have recognized the damaging effects of air pollution on human health? Without Walter Winchell speaking those forty words that alerted the nation to the deaths in Donora, who knows how quickly Truman and Eisenhower would have acted to prompt the steel, petroleum, paper, and glass industries to change their practices?

For that matter, without Rongaus, Koehler, Roth, and the other physicians in town giving injections and other drugs to treat deathly ill residents that weekend; without Schempp, Davis, and the other firefighters crawling through town with their oxygen packs, giving lifesaving oxygen to sick residents; and without the Schwerhas, Chambons, Stacks, and Uhriniaks of the town pulling together in the face of a major disaster, who knows how many more people would have died?

29

FIGHTING THE
GOOD FIGHT

AN AUTO-FREIGHT CLERK NAMED LEROY C. LE GWIN READ IN A WILM-
ington, North Carolina, newspaper about the deadly smog that had just
occurred and immediately contacted his friend, William Gillies Broadfoot
Jr., president of the Wilmington Junior Chamber of Commerce, or Jaycees. Broadfoot had been a heroic World War II fighter pilot, receiving
not only the Silver Star and Air Medal but also the Distinguished Flying
Cross, British Distinguished Flying Cross, and Chinese Distinguished
Flying Cross.

Le Gwin called Broadfoot with an unusual proposal, to fly ill residents
of Donora to Wrightsville Beach, near Wilmington, for rest and recovery. Wrightsville Beach possessed an abundance of sunshine and fresh air,
things the people in Donora and Webster hadn't experienced for several
sickening days. Could the Jaycees bring some of the residents to the beach
as a gesture of good will?

Broadfoot loved the idea and went to work. By November 2, 1948,
he had secured approval from the Wilmington and Wrightsville Beach
branches of the Jaycees to host guests from a town 430 miles away. August

Chambon must have felt overwhelmed with gratitude at the invitation. He accepted on behalf of his town and set about asking Donora's physicians to provide a list of patients who could benefit most from clean air and rest. Just over two weeks later forty valley residents were transported to Allegheny County Airport in West Mifflin, Pennsylvania, where they boarded an airplane with "Good WILLmington Mission" painted on the side. It was a first-ever flight for many of the residents.

Among the guests was Russell Davis, assistant fire chief, who had stumbled blindly through the smog to bring oxygen to sick townspeople. In the process he had become ill himself and too exhausted to continue. All but two of the guests hailed from Donora. The remaining two lived in Webster: Alice Ward and Kay (possibly Kaye) Weir, who, at twenty-six, was the youngest. Twenty-three men and seventeen women made the trip under the auspices of Regina Dougert, a Donora nurse. When the guests arrived in Wilmington they were met with flowers and photographers from local papers. The hosts provided dinners, baked goods, ice cream, and newspapers. They also offered laundry and dry cleaning services and even haircuts to their out-of-town visitors. The visitors, for their part, rested at the beach, visited a nearby plantation, enjoyed the genuine hospitality of Wilmington and Wrightsville Beach residents, and breathed in cool, clean, healing ocean air.

Stanley Michael Wazny, a forty-one-year-old laborer at the Zinc Works, was grateful to be one of the people chosen for the trip. "Two doctors gave me up," Wazny explained. "They told me I would have to leave Donora, so this trip couldn't have come at a better time."

Sixty-nine-year-old Alice Ward was also delighted to be a guest. Ward, an immigrant from Wales, had already crossed the Atlantic once by air and nine times by sea, so a short flight to Wilmington seemed rather effortless. "I've earned my living with my own two hands all my life," said Ward, "and it's the first time I've ever got anything for nothing. I'm going to make the most of it."

Back in the Monongahela Valley, anger was building. Doc Bill Rongaus was angry, to be sure, but so were a growing number of loved ones of the smog's victims. Over the coming year several area families came forward with lawsuits against US Steel, arguing that the steel mills and zinc smelter were the proximate cause of the death of their family member. Many of the families decided to work through the legal firm of Margiotti and Casey, a Pittsburgh firm, with attorney Marvin Power handling the cases.

Marvin Dunbar Power, at fifty-five, was a highly experienced attorney by the time he took on the smog cases. He had handled many suits brought against corporations for injuries or deaths from faulty equipment, lack of meeting minimum standards, and the like. Power was born in Illinois but had lived in Dormont, just a few miles southwest of Pittsburgh, for close to thirty years. Power almost certainly realized the scope of difficulties he and his clients would face in confronting US Steel's extensive network of lawyers and its profound financial backbone. He nevertheless chose to pursue the case. He might have wanted to take the case all the way to trial but probably assumed that it wouldn't make it, that it would be settled before then.

If so, he was right. Power faced the same firm that attorney Pinkasiewicz had faced in the Gliwa suit against US Steel years before: Reed, Smith, Shaw and McClay, this time with Charles E. Kenworthey and Carl E. Glock Jr., serving as co-counsel. Power's most compelling case was the one brought by John Gnora, husband of Susan Gnora, the smog's tenth victim. Power filed at least nine other suits as well, for people who died during the smog or during the coming year.

The principal witnesses in the Gnora case were John Gnora, a few members of his family, and William Rongaus, who had treated Susan when she became ill. Initial charges were brought in September 1949, but the case didn't reach the deposition stage until the following September. Gnora sued US Steel for an amount sufficient to compensate the family not just for Susan's death but also for damage to the Gnora home from noxious fumes emitted by the metal plants. The suit contended that the Donora Zinc Works emitted "certain noxious fumes, poisonous gases, solids, and chemicals into the atmosphere," and that those elements constituted a "nuisance." It further charged that even though US Steel knew about those chemicals it "willfully, wantonly, and maliciously continued such nuisance and failed to take any corrective action whatever."

The suit, in its sixteenth and final charge, claimed that because of "exposure to the noxious fumes, gases, chemicals, and solids emitted by defendant's plant and zinc works, [Susan Gnora] was caused to be sick, sore, and disabled, [which] resulted in her death on October 30, 1948."

US Steel's attorney Kenworthey responded to the suit with two basic defenses, the same defenses he would use for each of the other cases. First, US Steel admitted that its plants did produce zinc, sulfuric acid, and "small quantities of cadmium" and that Susan Gnora did indeed die on October 30, 1948. However, Kenworthey denied "each and every other allegation contained in the complaint." And anyway, Kenworthey continued in his other defense, even if there was a "serious risk" to Gnora and her family, it was they who "voluntarily assumed that risk."

Kenworthey further requested answers to a set of interrogatories, questions from one party to the other that help determine essential facts of the case and which facts would be presented at trial. In the Gnora case, Kenworthey threw more than thirty interrogatories at Power and his clients, including these:

- What was the date of each other illness or physical injury of decedent during the ten years preceding her death?

- How many of such trees and shrubs [on the property] are claimed to have been damaged or destroyed by defendant's negligence?

- For each such repair job [to the house] give the nature of the work done, the materials used, the cost including materials, labor and any other charges, and the name and address of the person to whom the cost was paid.

Power would have asked John Gnora for help in answering all of those rather intimidating questions. How this uneducated laborer from Slovenia who worked in dangerous conditions at a metal factory, who spoke little English, and who had just lost his wife of forty-one years could have coped with such a flurry of queries cannot be comprehended, never mind a later interrogatory that demanded receipts for all of those repairs.

The drawn-out legal process proved too much to bear, and John Gnora finally settled the following year. His suit sought a total of $80,000 in damages, an amount worth more than $925,000 in 2022. In the end Gnora, and each of the plaintiffs in the other suits, decided to settle out of court. Gnora received just $5,000, out of which came $1,666.67 for his attorney and $277.77 for each of his eight children, a pittance by any standard. By comparison a brand-new RCA television in 1951, the year the Gnora suit was settled, cost $269. For himself John Gnora was left with just $1,111.11 as compensation for the death of his beloved wife.

Gnora had put his trust into Marvin Power's hands. Whether Power served that trust well can be debated. No doubt he believed, however, that after running headlong into US Steel's wall of inexhaustible funds, that the settlement he had gained for his clients was the best he could have hoped for.

US Steel was also sued by a more organized group of Websterites called the Society for Better Living, initially formed in the 1930s by steel worker Stephen "Beanie" Huhra. The group fought valiantly against the Donora mills, particularly the Zinc Works, and protested frequently. Huhra often rode a donkey-pulled buggy in Donora parades in protest of the zinc smelter. Some historians consider the group one of the first true grass-roots organizations in the national environmental movement. After the smog the group, headed at that point by prominent Webster resident Abe Celapino, filed lawsuits against the mills. Like Gliwa and Pinkasiewicz, though, Celapino ran into the US Steel buzzsaw and eventually lost the case.

In total about 130 lawsuits were filed against US Steel in the wake of the smog disaster. The suits claimed total compensation of $4,643,000 for the victims and their families. All were settled out of court for a total of $235,000, just 5 percent of the amount sought.

Environmental researchers might have hoped that a Public Health Service study on the tenth anniversary of the smog disaster would provide compelling support for even stronger air pollution controls. Unfortunately it was not to be.

The report prepared by Antonio Ciocco, the first chair of the Department of Biostatistics at the University of Pittsburgh, and Donovan J. Thompson, professor of biostatistics there, collected mortality data on nearly every individual studied by the Public Health Service immediately after the smog. The men tried to answer the question, "Did illness experience vary with exposure to pollutants?"

Differences in mortality among smog victims did exist, they found, and depended greatly on whether they suffered at the time from a chronic heart or lung condition. However, they couldn't say with certainty that those differences were directly due to the leftover effects of the smog or some other condition. For instance, a great many residents at the time smoked cigarettes, lived in homes heated by coal, or both. Ciocco and Thompson were unable to separate the effects of specific toxins in those kinds of smoke from the same toxins emitted from the mills.

The researchers acknowledged that their studies were "admittedly incomplete and of necessity crude." In the end they, too, failed to point any scientific fingers at the mills. "Although the Donora data do point to some effects on the cardiorespiratory system," read the anniversary report, "they do not provide the means for relating particular pollutants to specific symptoms."

Like the Townsend report in 1949 and the Meuse Commission findings well before that, not enough data existed to definitively tie toxic factory emissions to rates of illness and death in individuals exposed to those emissions. The key word there is "definitive." A quick look at the death certificates for smog victims would yield clues. Eleven of the twenty-one victims died from heart disease, usually noted as myocarditis, an inflammation of the heart. It seems likely that all twenty-one people suffered some degree of heart disease, though it might not have been diagnosed yet.

Scientists have long identified a link between air pollution and heart disease, but they have been unsure of its exact cause. Now they know. In 2016 a ten-year-long study called the Multi-Ethnic Study of Atherosclerosis and Air Pollution identified a key link between air pollution and heart disease. Researchers at the University of Washington examined the thickness of artery walls in test subjects, as well as each subject's exposure to a set of common pollutants: carbon oxides (mon-, di-, and trioxide), fine particulate matter, and black carbon, a type of fine particulate matter more commonly known as soot. The results indicated that air pollution accelerates the hardening of cardiac arteries, a condition that typically occurs along with plaque buildup in the arteries. Together those two conditions lead to more deaths per year worldwide than any other disease.

Air pollution has also been clearly linked to a wide variety of other conditions, including low birth weight; delays in the development of a child's brain; a worsening of diabetes; numerous chronic respiratory diseases, including asthma, pulmonary sarcoidosis, pneumoconiosis, and obstructive pulmonary disease; as well as a variety of cancers, including and especially lung cancer, but also leukemia and breast, liver, and pancreatic cancers.

Older adults are especially vulnerable to the pernicious effects of air pollution. A team of researchers at the Harvard School of Public Health found in 2021 that even low levels of air pollutants in the air can increase the risk of heart and lung diseases in the elderly. The study's lead author, Mahdieh Danesh Yazdi, urged doctors and their patients to advocate for cleaner air and to "apply pressure on public officials to control the sources of pollution and improve the air we all breathe."

Yazdi continued, "Even if air pollution can't be fully mitigated, we should strive to do better. Levels of pollutants now considered safe can still have harmful effects and result in bad outcomes."

The Donora smog might have served as a catalyst for the Harvard study and thousands like it, but Mon Valley residents still wrestled with what they perceived as their wretched place in world history. "Twenty-plus people dying in your town over the course of a weekend," explained historian Brian Charlton, "is not something you put on your chamber of commerce list of accomplishments." Donorans felt ashamed of their town and thought that residents in surrounding communities, particularly Monongahela, Charleroi, and Belle Vernon, considered Donora a disgrace, a town that had dishonored the whole valley. Bitterness existed for many years, even among town leaders. Charlton remembered former Donora mayor John Lignelli once telling then president of the Donora Historical Society Ruth Wickerham Miller, "Your smog is just an albatross around our neck."

Not even historians paid much attention to the disaster until the early 1990s, after President George H. W. Bush signed the Clean Air Act Amendment of 1990. The amendment expanded the act's focus and directed it toward implementing innovative technologies, reducing acid rain, and opening the markets to alternative fuels. Said Charlton, "That's when historians began to mark Donora as the place where the modern environmental movement started."

The town has since recognized the unique role its tragedy played in the significantly cleaner air enjoyed by the nation today. For decades the town's slogan has been "Next to Yours the Best Town in the USA." That slogan remains a source of pride for Donorans, who believe their town is "the best" but who also graciously concede that, well, maybe "your" hometown is pretty good too.

The majority of Donorans today embrace their history. They point with honor to the town's main attraction, the Donora Historical Society and Smog Museum, the keeper of the tragedy's flame and the principal storyteller of the town's many celebrated offspring. The museum has its own slogan, one birthed in tragedy and instilled with pride. The slogan parses the town's true legacy into just four words:

CLEAN AIR STARTED HERE

EPILOGUE

AS THE 1950S PASSED, FEWER AND FEWER ORDERS CAME IN TO THE STEEL and zinc facilities in Donora. The drop in orders had nothing to do with the smog and everything to do with the age of the equipment and the cost of maintaining it. By 1950 the Zinc Works was forty-five years old and the steel mills and furnaces a full fifty years old, and none of the mills had seen more than a hint of modernization. Newer and far more efficient processes existed by then, particularly in zinc production. US Steel finally said enough and closed the Zinc Works in 1957. The steel mills made it another decade before shutting down for good in 1967.

The air over the Mon Valley cleared significantly after that, and vegetation grew anew. Grass, trees, shrubs, and bushes began to thrive throughout the valley. The hill above Webster, once a muddy, rutted hill, now is carpeted in forest.

Air in the valley remains far cleaner than it has been in more than 150 years, yet air pollution remains a sizable problem. In 2018 more than 4,800 Pennsylvanians died prematurely from air pollution. Where the state's economy once depended on the production of steel and other metals, it now depends on the production of energy for an increasingly electronically fueled populace. A dominant amount of the state's air pollution currently stems from power generation and vehicular traffic. Pennsylvania's air quality isn't the nation's worst, though. That murky honor goes to California, with Texas, Washington, and Oregon rounding out the top five.

As unhealthy as US air can be, it is far cleaner than air in other parts of the world. People living in Rome, Sarajevo, areas of Mongolia, and regions in Iran, Afghanistan, and Pakistan suffer extreme air pollution on an almost daily basis. Worst of all is India, with an air quality index two and a half times that of its nearest "competitor," Kosovo. Of all the cities in India, New Dehli is far and away the most polluted.

Furkan Latif Khan, an NPR producer in New Delhi, said, "The first thing I do every morning is open my balcony door and check the air outside. If I spot the blue sky, I'm overjoyed and draw in a deep breath. But many days, the air has a dusty, burnt taste. I make a mental note not to forget my air filter mask before I leave the house." Khan knows well the damaging effects of air pollution. "Short- and long-term exposure to air pollution," she said, "has a direct association to respiratory and cardiovascular diseases, eye irritation, skin diseases, and cancer."

Living as she does in one of the world's most polluted cities, she can hardly escape the air. Khan wonders, as millions of other Delhi residents surely do as well, how all of the emissions from industries and power plants, exhaust from cars and trucks, dirt and dust from construction sites, and smoke and fumes from distant fires will affect her own health. "Every day," she said, "I wonder how many years of my life I am giving up, breathing in New Delhi."

Hopeful signs exist in the fight to clean the world's air. Many nations, and the larger cities within them, have taken substantive steps to lower emission levels of pollutants over the last few decades. Residents of London today, for instance, breathe much cleaner air than they did in 1952, when the Great Smog of London settled over the city for five days in December.

The city had been dealing with several weeks of unrelenting cold at the time. Most of the homes and businesses were being heated by coal-burning furnaces, a great many of which used a cheap fuel called nutty slack, consisting of coal dust (slack) and morsels of coal (nuts). Nutty slack was notoriously inefficient, and furnaces required large amounts to provide even close to adequate heating. Nutty slack also emitted huge amounts of carbon monoxide and sulfur dioxide. When a prolonged temperature inversion struck, those gases found no escape into the atmosphere, and the crowds of people on London streets breathed increasingly toxic air.

The US Public Health Service had sent the Townsend report to the British government shortly after it was published in 1949. Tragically, London failed to react to the smog as swiftly as Donora had, and in fact reacted hardly at all. "A city hardened by war," wrote Kate Winkler Dawson, author of *Death in the Air*, "still believed the fog was simply a prolonged peasouper, just another byproduct of living in London." Where Donorans had sprinted to help others all over town, Londoners barely looked up from their afternoon tea. "They [Donorans] set up a triage center in the community center," said Dawson. "It was a concerted effort from everyone in the town to try to get through this. And they recognized it as a deadly disaster."

Dawson explained that in London "the panic didn't even come until months later when the death tolls came out and they realized that thousands of people died. And even then there wasn't panic. So in Donora, there's a museum dedicated to the event, and there's nothing like that in London."

Even without a museum dedicated to London's eight thousand to twelve thousand smog victims, the British government has, over time, successfully slashed its level of air pollution, particularly in its capital. Since 1935 the average concentration of suspended particulate matter in the air over London has steadily decreased, from 409 micrograms per cubic meter to just sixteen in 2016. A major part of that drop stemmed from the United Kingdom's first clean air act, passed in 1956 and expanded in 1968. London still suffers severe smog, but nowhere near as bad as during that brutal winter of 1952.

Survivors of Donora's smog understand well what those Londoners must have felt. Comparatively few of those residents survive today, however. Some have moved; many more have passed on.

After bulldozing an entire town into existence in 1900, William Donner left Donora far behind. A few years after becoming president of Cambria Steel in 1902, Donner assumed the chairmanship of Pennsylvania Steel Company. Then, at some point during the mid-1910s, Donner left Pittsburgh and moved to a prominent brick building at the corner of Pine and Eighteenth in Philadelphia, the city where his wife, Dora, had grown up. Donner purchased the New York State Steel Company in 1915, renaming it Donner Steel Company. He remained a leader in the steel industry until after his son, Joseph, was diagnosed with cancer in early 1929.

Joseph Donner, born July 15, 1894, graduated from Princeton University in 1917 and later married Carol Elting, daughter of Howard and Florence Elting, a prominent Chicago couple. Joseph began feeling unwell shortly before Christmas 1928, when he and his wife were living in Buffalo. Joseph's physician, Theodore Wright, could find nothing wrong with Joseph after his first examination in December. Nor could physicians in Philadelphia, nor those at Johns Hopkins Hospital in Baltimore. Without a clear diagnosis, the doctors advised Joseph to ease back on work—he was serving as vice president of Donner Steel at the time—and rest. His health did not improve.

The following May Joseph and Wright were riding an elevator at the Saturn Club, one of Buffalo's premier private clubs, when Wright noticed an odd lump in Joseph's neck. "How are you feeling, Joseph?" The doctor asked. "Do you have a cold or sore throat?"

"No," Donner replied, "why?"

Wright said that he had spotted a "slight swelling" in one of the lymph nodes in Joseph's throat and that he should stop by his office after lunch, which he did. Wright made some pretext to take a biopsy of the node and was perhaps not surprised when under the microscope he found cancerous cells.

Joseph's brother, Robert, gave their father the bad news. "Father, we know what is wrong with Joe now," Robert said. "I have not wanted to worry you."

"What is it?" the elder Donner asked.

"A tumor."

"Is it malignant?"

Robert responded simply, "Yes."

Joseph succumbed to his disease a few months later, at age thirty-five, on November 9, 1929. An autopsy showed that Joseph had died of lung cancer, a disease now conclusively linked to the very pollutants emitted by his father's steel mills.

The news surely stunned the elder Donner, though by the time he wrote his autobiography, late in his life, he had gained enough perspective to afford the event a typically clinical description. Donner wrote only, "It was a metastasis of a malignant tumor."

Donner wrote nothing of his reaction to Joseph's death. He was such a private man that he could not bring himself to share his feelings openly, even about such a heart-rending moment. Nor did he mention in his diary or elsewhere his reaction to another sad event in the family a few years later. Elizabeth Browning Donner, the second child of William and Dora, married Elliott Roosevelt in 1932. Elliott was the son of then-governor of New York and soon-to-be president Franklin Delano Roosevelt. Eighteen months and one child later the pair divorced. Elizabeth charged Elliott with "extreme cruelty" and said that "her temperament and that of her spouse differed so greatly that further life together was out of the question." Five days later Elliott married Ruth Googins, who later divorced him for "unkind, harsh and tyrannical conduct." William Donner might have been filled with rage at Elliott's treatment of his daughter, but if he was, he seems never to have mentioned it publicly.

Joseph's illness changed the trajectory of Donner's life. That summer, as Joseph was undergoing radiation treatment at Johns Hopkins, his father began planning a foundation to support cancer research. "It became apparent to me," wrote Donner, "that very little was known about cancer, and I decided that someday I would give a substantial amount for cancer research." He sold Donner Steel Company that fall and with an initial fund of $2 million established the International Cancer Research Foundation in Philadelphia to underwrite studies on cancer.

Donner retired in 1937 and, after the end of World War II, purchased a villa in Montreux, Switzerland, a picturesque settlement on the shores of Lake Geneva. He and Dora often visited Montreux during their later years. They also spent a great deal of time in Montreal, Canada; so much so that Donner would contribute over the years hundreds of thousands of dollars to McGill-Queen's University, including one gift of $232,355 to construct the Donner Building for Medical Research. He closed his cancer research foundation in Philadelphia after the war and used its assets to start a new charitable endeavor, the William H. Donner Foundation, currently headquartered in Tarrytown, New York.

Becoming increasingly unhappy with US tax rates Donner created the Donner Canadian Foundation in 1950. That philanthropic, fiercely conservative foundation remains active today and is headquartered in Montreal.

Three years later Donner, by then a widower, became seriously ill while in Switzerland. He flew to Montreal and was immediately admitted to Royal Victoria Hospital. He died a week later, on November 3, 1953. He was eighty-nine.

Zinc mill superintendent M. M. Neale remained in his position until he retired in 1951 at age sixty-five. It appears that none of the mill's leaders were fired for their role in the smog, and it seems doubtful any were even temporarily suspended. For that to have happened, US Steel would have had to admit it was in some way wrong, an utterly unpalatable option. Superintendent Neale was actually feted at his retirement by members of the Zinc Works union, USA-CIO Local 1757. The heading on the celebration's program read, "A Regretful Loss to Donora." Neale lived out his days in the bucolic farm-and-forested community of Heathsville, Virginia, a hamlet just a few miles from the Chesapeake Bay. Neale died of a heart attack on December 11, 1968, at age eighty-two.

Rudolph Schwerha, the mortician who, together with an unknown assistant, braved the deep fog repeatedly to gather the bodies of smog victims, retired in 1963. He had provided mortuary services for Mon Valley residents for thirty-five years. He served on the Donora school board for sixteen years and was four times its president. After retiring Schwerha moved first to Winsted, Connecticut, then in 1984 to his final home in Williamsburg, Virginia, where he died on October 7, 1985, at age eighty-four.

Donora's fire chief, John Volk, guided his department for twenty-seven years before retiring in 1973. He died in October 1981, just shy of his eightieth birthday. Volk's assistant, Russell Davis, was still an active firefighter when a sudden stroke felled him on November 26, 1954. He was just forty-two.

Helen Stack, the harried office assistant for Koehler and Roth, decided to stay in health care after the smog. She graduated from the Modern School of Practical Nursing in Pittsburgh in 1963. She outlived both physicians and many smog survivors, dying at eighty-nine on July 10, 2010.

Ralph Koehler practiced medicine in Donora for many years. His son, Donald, also became a physician and joined his father in practice in May 1958. The elder Koehler survived diabetes, two successive heart attacks, and a stroke before perishing from heart failure on April 30, 1967. He was sixty-seven.

The cigar-loving Edward Roth continued to practice medicine until forced to retire. Roth, serving a term as head of the medical staff at Charleroi-Monessen Hospital in 1965, suffered a debilitating heart attack. "He would've died if he hadn't been in the hospital at the time," said nephew Jerry Harris. Roth halted his practice in Donora, telling his family, "I've got to get out of here." He and Sally moved to Phoenix to rest. Feeling somewhat stronger and still craving the need to practice his life's work, Roth joined the infirmary at Arizona State University, treating students and, in his spare time, conducting employment examinations for Maricopa County Hospital, now part of Valleywise Health.

While driving home on the afternoon of January 23, 1978, Roth suddenly slumped over the wheel of his car and crashed. He was seventy years old. He most likely died at the scene from a massive heart attack, though his obituary lists him as having died at nearby Saint Joseph's Hospital. His adoring Sally would live twenty-seven more years before dying at age ninety-one on August 27, 2005.

Doc Bill Rongaus spent the rest of his career in Donora. He served on the medical staff at Charleroi-Monessen Hospital, later renamed the Monongahela Valley Hospital, and served at least once as its chief of staff. In 1983 he received the hospital's most prestigious award, the Edward J. Protin Memorial Award, for "significant contributions to the hospital" and for his "qualities of leadership, loyalty, compassion, and empathy." Rongaus remained a generous citizen, caring physician, loving father, and intense protector of the underdog throughout his life. He developed leukemia in his later years and died at his home hospital on March 1, 2001, at eighty-six.

Bill Schempp became the unofficial spokesman for the Donora Fire Company in the immediate aftermath of the smog and for years thereafter. His likability and ebullience made him a natural for talking with reporters and newscasters. He served with the fire company for many years and worked his way up to assistant fire chief. He even owned a fire truck, a 1947 American LaFrance, the so-called Rolls-Royce of fire engines. He was a longtime counselor for the Boy Scouts of America, president of the county's postmasters association, and the Washington County Fireman of the Year in 2008.

As Schempp aged his health started to decline. He developed arthritis and macular degeneration, neither of which stopped him from getting around town. Schempp was a stoic man, never complaining about the constant pain he suffered. His daughter, Annie, heard him complain only once about pain, when his arthritis flared. Annie asked him, "'Daddy, why don't you take some aspirin?' He said, 'I don't want to get hooked.'"

Bill and Gladys Schempp were inseparable, one of those couples everyone knew, liked, and respected. They adored each other and the life they had made in little Donora, so when Bill died at ninety-one, on December 14, 2009, after sixty-seven years of marriage, Gladys wasn't far behind. She died just twenty-nine days later, at age ninety.

Few traces remain of the mill complex in Donora. Most of the buildings that once produced staggering amounts of steel and zinc and released an incalculable amount of toxins into the air over their half century of use have been demolished. The land that held that complex, which once consisted of hundreds of buildings and nearly thirty miles of railroad track, now houses a nondescript industrial area with large open spaces of dirt, weeds, and leftover slag.

One of the original buildings still standing is the former Zink Works hospital. Physicians and nurses working in that small two-story brick building treated thousands of workers and an enormous variety of injuries and medical conditions over the years. To walk by the building today is to be reminded of the souls lost in the forging of America's greatest icons of progress. It is to be reminded of the workers who made the metals that supported two world wars and helped to reinvigorate America after the Great Depression.

It is to be reminded of the brave men and women who worked tirelessly at the plants and under the most arduous and perilous conditions imaginable. It is to be reminded of the families of those workers, families who persevered through unfathomable grief and interminable hardship to support their loved ones.

And it is to begin, but only begin, to understand the immense contribution made to the nation's prosperity by the tens of thousands of immigrants over the decades who came to Donora for a better life: immigrants who worked, played, lived, and died together in a steep-walled valley of a horseshoe-shaped curve along a north-flowing river during the first half of the twentieth century.

ACKNOWLEDGMENTS

WRITING A BOOK IS BOTH A SOLITARY ENDEAVOR AND ONE FILLED WITH unexpected friends, visits to new places, and exposure to a potpourri of new faces, voices, and outlooks. I have enjoyed my time with every person I met and am profoundly grateful for the help I received since beginning my adventure with Donora.

I would like to thank some of the many people I spoke with, including and especially the following (in alphabetical order): Ken Barbao; Graham Davis, who provided a wealth of information on firefighter Russell Davis; Eddie Diaz; the delightful Jerry and Lynn Harris; John Hepple; Irene Louise Hilaire and the late, lovely Rose Marie Iiams, my first interviewees for the book; Karl Jackson, owner of the House of Jackson Funeral Home, formerly owned by Rudolph Schwerha; Edith Jericho; Les and Bill Kilduff, grandsons of Milton Mercer Neale; Scott Kirkwood; Rick Lewis, who provided wonderful background information on the Rodriguez clan; the marvelous Patricia Ann McCarthy Dugan; Sidney Mishkin, who selflessly provided his memoir to me; Carol Ochadleus; Ron Paglia; Dave Papak; Dimitri Petro, whose memory marvels me still; his brother, Paul Petro; Bonnie Poklemba; Marvin Preston for his honesty; Henry Pykosh; Jean Rose; the marvelous Bonnie and Shirley Rozik; James Sawa, footballer Stanley's son; Yugan Talovich; the late, gracious Lana Vrabel and her daughter, Cheryl Munaretto; and the renowned judge Reggie Walton, who kindly spoke with me during what must have been an unusually hectic time for him.

Researching a nonfiction book today might be easier than ever with so much material online, but the value of hands-on research and the expertise of professional researchers cannot be underestimated. I am indebted to all the help I received from Grace Schultz, archivist at National Archives at Philadelphia; Michael Dabrishus, assistant university librarian, University of Pittsburgh Hillman Library; David Grinnell, coordinator of archives and manuscripts, University of Pittsburgh; the always helpful journalist and blogger Scott Beveridge; Lauren Kifer, communications director for Mon Valley Hospital; Erin Lavin, registered dietitian and certified nutrition support clinician at Tulane University; and the librarians at Donora Public Library. A special thank you to Eric S. Lidje, director of the Rauh Jewish Archives at Heinz History Center in Pittsburgh, and Dan Zyglowicz of the Archives and Special Collections of California University of Pennsylvania, both of whom answered my repeated queries with patience and interest.

Meteorology and the zinc- and steelmaking industries were foreign to me when I started, but several marvelous individuals assisted me greatly in understanding those highly complex processes. I am most fortunate in having had access to several remarkable educators at Carnegie Mellon University, including Joel Tarr, Richard S. Caliguiri University Professor of History; Cliff Davidson, professor of civil and environmental engineering; and Neil Donahue, professor of chemical engineering, chemistry, and engineering. It was my extreme pleasure to spend time with the energetic and brilliant George Leikauf, professor of environmental and occupation health for the Graduate School of Public Health at the University of Pittsburgh, and to be able to pick the brain of the wonderful Donald I. Bleiwas, retired geologist for the US Geological Survey and coauthor of the enormously helpful report "Historical Zinc Smelting in New Jersey, Pennsylvania, Virginia, West Virginia, and Washington, D.C., with Estimates of Atmospheric Zinc Emissions and Other Materials." I am equally grateful to Ronald Baraff and his team at the monumental Rivers of Steel National Heritage Site (absolutely worth a visit); P. Chris Pistorius, POSCO Professor of Materials Science and Engineering at Carnegie Mellon, who answered every question I asked about weather conditions in the Mon Valley; and Art Larvey, a dedicated, exceptional, retired chemical engineer.

Donora's founder, William Henry Donner, was such a private man that little public information exists about his life. I am therefore supremely grateful to have been able to use Donner's privately published memoir, sent from his great-grandson Ben Denckla. I am similarly indebted to Donner's relatives Curt Winsor and Timothy Donner, both of whom proved open, gracious to queries, and supportive about this book.

I am grateful to Lynn Page Snyder for her doctoral dissertation on the Donora smog, a constant companion during my writing, and to Devra Lee Davis, an environmental epidemiologist and founder and president of Environmental Health Trust, a nonprofit environmental research organization. Davis wrote *When Smoke Ran like Water: Tales of Environmental Deception and the Battle against Pollution*, which offered a chapter on her childhood in Donora and the smog that beset it, as well as an in-depth look at land and water pollution throughout the nation.

I am grateful to David Lonich for his knowledge of Donora's history and to the delightful, late Charles Stacey for his voluminous memory and his kindness in imparting his knowledge and wisdom.

The marvelous Annie Schempp, daughter of Bill and Gladys Schempp; the inimitable Nancy Rongaus Cherney, daughter of Dr. William Rongaus; and the remarkable Jerry Harris, nephew of Dr. Edward Roth, each provided an enormous amount of information about their heroic dads and did so without hesitation. Their kindnesses, humor, and dedication to an accurate retelling of those dark days can never be repaid.

I am thankful to Sandy Crooms, former editorial director at the University of Pittsburgh Press, for taking a chance on a guy like me, and to Abby Collier, who competently and lovingly saw the project through to fruition. Thank you as well to my agent, Amanda Jain, of Bookends Literary Agency. I knew almost immediately that the friendly, diverse, and talented Bookends crew was the place for me, and Amanda my kind of agent.

This book would never have been completed—and indeed would not have been started—without the cooperation and ardent support of Brian Charlton, historian of the Donora Historical Society. Brian believed in the book and helped me not only with basic historical information about the smog but also with his perspectives as a historian. His feedback proved invaluable throughout, and I am hugely grateful to him.

I am forever indebted to Mark Pawelec, an intrepid soul who aided my research in innumerable ways. He reviewed every chapter and was instrumental in helping to ensure accuracy and completeness about all things Donora. Mark gave freely of his time, as he does for anyone asking about

Donora or its history, and I know he will continue to do so long after this book has found the bottom of the dollar-sale bin. I am honored and grateful that he gave so much of his time and effort to me. If errors exist in the book, they are my fault entirely.

My greatest thanks, though, go to my beautiful and hugely supportive wife, who has been with me through the highs and lows of writing and getting a book published. I thank her, and I will love her to my dying day.

APPENDIX

COMPLETE VICTIM DATA

This section presents detailed information on all victims.

Determining Who Was a Victim

The most difficult obstacles I faced early in writing this book were in determining the exact number of people who died in the smog. Written accounts in newspapers, magazines, and books—not only contemporary ones but also those written decades later—varied substantially in their victim counts. Many newspaper accounts at the time listed the final tally of deaths at twenty; others listed seventeen, eighteen, occasionally nineteen. A large bronze plaque presented to the town in 2008 lists the names of twenty-seven victims. To produce the most accurate list possible I developed a set of criteria for determining who should be counted among the victims and who should not.

First, I chose the smog's start and end dates as determined by historians at the Donora Historical Society and supported by national weather reports—October 26 to the afternoon of October 31. Other start dates, including one used by the federal government, are later, generally October 27. No end date I found extended beyond October 31.

Second, the death of a victim must have been directly associated with the smog, either through a physician's or coroner's reported cause of death or through contemporaneous, reasonably reliable third-person accounts. It is here that many people whose relatives died "in" or "from" the smog have not been counted. For example, although the father of baseball legend Stan Musial, a Donora native, is often said to have died in the smog, Lukasz Musial died from a stroke on December 19, 1948. The death certificate provides no information that might indicate a death directly related to the smog. In addition, Musial had suffered a stroke four years before and suffered from a heart condition called atrial fibrillation, both of which made him significantly more likely to suffer another, possibly fatal, stroke. I have little doubt that the smog affected Musial's overall health, as it did so many others, but without relevant data I have considered him, and many other people in similar circumstances, to have been one of the thousands of people who were sickened *by* the smog but were not direct victims *of* it.

Third, the death certificate, if available, must indicate one or more cardiopulmonary disorders as a primary or secondary cause of death. Typical disorders for such an event would include asthma, bronchitis, pneumonia, or heart failure, sometimes called cardiac asthma in the medical nomenclature of the time.

Fourth, a victim must have died either during the smog or within forty days thereafter, the time delay used by air pollution epidemiologists. George Leikauf, professor of environmental and occupation health at the University of Pittsburgh, explained: "Typically in epidemiological studies there is a forty-day lag period in mortality included after an air pollution episode. Many mortalities occur during or immediately after the episode, but many occur days or weeks later. Deaths that occur within forty days of an episode can be attributed to the episode." Applying that lag period to the Donora disaster indicated that two people, in addition to the nineteen who died during the smog, should be counted as victims: Thomas Amos Short, who died November 8, and George Hvizdak, who died in early December.

Finally, individuals living outside the immediate environs of Donora and Webster have not been considered victims unless it could be determined with a high degree of confidence that they spent a significant amount of time in Donora or Webster during the smog and if they also met all other criteria. As it turned out, I found no such victims.

Where Each Victim Lived

This table provides street addresses, where possible, and the town or area where each victim lived. The race of each victim is also presented and helps to show, to an extent, the subtle stratification that existed in the valley. Only one Black resident, Perry Stevens, lived in Donora, and that was in a boarding house near Cement City. All four of the other Black victims lived across the river, where rents and home prices tended to be considerably lower than in Donora.

VICTIM	RACE	STREET	TOWN OR AREA
Ivan Ceh	White	Rear 464 Fifth Street	Donora
Barbara Chinchar	White	50 Watkins Avenue	Donora
Taylor Circle	White	725 Heslep Avenue	Donora
John Cunningham	Black	322 Tenth Street	Donora
Bernardo Di Sanza	White	337 Third Street	Donora
Michael Dorincz	White	532 Ohio Street	Donora
William Gardiner	White	725 McKean Avenue	Donora
Susan Gnora	White	Donora Place Plan	Donora
Milton Hall	White	East Donora Road	Rostraver
Emma Hobbs	Black	Unknown	Webster
Ignace Hollowiti	White	83 Allen Plan	Donora
George Hvizdak	White	Unknown	Sunnyside
Jane Kirkwood	White	121 Ida Avenue	Donora
Marcel Kraska	White	715 Fourth Street	Donora
Andrew Odelga	White	450 Eighth Street	Donora
Ida Orr	Black	Unknown	Fellsburg
Thomas Short	White	PO Box 127	Webster
Peter Starcovich	White	Unknown	Wilco Hill
Perry Stevens	Black	88 Bank Street	Donora
Sawka Trubalis	White	438 Sixth Street	Donora
John West	Black	Unknown	Forward

When Each Victim Died

This table shows each victim's age and the day and time of death noted on each person's death certificate.

NO.	NAME	AGE	DAY	TIME (ET)
1	Ivan Ceh	69	Friday, October 30	1:30 a.m.
2	Jane Kirkwood	67	Friday, October 30	2:00 a.m.

3	Peter Starcovich	67	Friday, October 30	2:30 a.m.
4	Ida Orr	58	Friday, October 30	3:00 a.m.
5	John Cunningham	63	Friday, October 30	4:55 a.m.
6	Andrew Odelga	69	Friday, October 30	6:00 a.m.
7	Ignace Hollowiti	64	Friday, October 30	6:45 a.m.
8	Emma Hobbs	55	Friday, October 30	7:30 a.m.
9	Perry Stevens	55	Friday, October 30	7:45 a.m.
10	Susan Gnora	62	Friday, October 30	8:30 a.m.
11	William Gardiner	66	Friday, October 30	10:30 a.m.
12	Marcel Kraska	65	Friday, October 30	11:45 a.m.
13	Milton Hall	52	Friday, October 30	1:45 p.m.
14	Michael Dorincz	84	Friday, October 30	3:00 p.m.
15	Bernardo Di Sanza	67	Friday, October 30	3:30 p.m.
16	Taylor Circle	81	Friday, October 30	4:00 p.m.
17	Barbara Chinchar	55	Friday, October 30	10:30 p.m.
18	John West	51	Saturday, October 31	5:00 a.m.
19	Sawka Trubalis	65	Saturday, October 31	12:10 p.m.
20	Thomas Short	81	Monday, November 8	5:05 a.m.
21	George Hvizdak	52	Wednesday, December 22	11:15 p.m.

Other Demographics

This section outlines key statistics on smog victims, including age range, gender, birthplace, cause of death as listed on the death certificate, and the cemetery where each victim is buried.

AGE RANGE		GENDER		BIRTHPLACE	
AGE	VICTIMS	Male	16	Foreign	14
50–55	6	Female	5	United States	6
56–60	1			Unknown	1
61–65	5				
66–70	6	**RACE**			
71–80	0	White	16		
> 80	3	Black	5		
TOTAL	21				

COMPLETE VICTIM DATA

LISTED CAUSE OF DEATH

Ivan Ceh	Acute cardiac dilatation, cardiac asthma, chronic myocarditis
Barbara Chinchar	Pulmonary tuberculosis for two years, acute tracheobronchitis
Taylor Circle	Chronic myocarditis for ten years, arteriosclerosis, bronchial asthma
John Cunningham	Cardiac asthma, unknown duration
Bernardo Di Sanza	Asthma
Michael Dorincz	Extreme edema of the lungs due to hypertrophy of the heart due to mitral valvular disease
William Gardiner	Cardiac failure due to asthma
Susan Gnora	Asthma
Milton Hall	Chronic myocarditis
Emma Hobbs	Status asthmaticus
Ignace Hollowiti	Pneumonia for ten days, tuberculosis
George Hvizdak	Chronic myocarditis, lung abscess nontuberculous
Jane Kirkwood	Asthma
Marcel Kraska	Asthma
Andrew Odelga	Bronchial asthma
Ida Orr	Acute myocarditis for one day, cardiac asthma for one year
Thomas Short	Asthmatic bronchitis for ten days, generalized arteriosclerosis
Peter Starcovich	Chronic myocarditis
Perry Stevens	Cardiac failure, asthma
Sawka Trubalis	Chronic myocarditis for ten years, acute tracheitis, bronchitis
John West	Bronchial asthma for two to four hours, acute tracheobronchitis for three hours, "abnormal weather conditions"

CEMETERY

Ivan Ceh	Saint Dominic
Barbara Chinchar	Saint Michaels
Taylor Circle	Monongahela
John Cunningham	Monongahela
Bernardo Di Sanza	Saint Dominic
Michael Dorincz	Saint Michaels
William Gardiner	Monongahela
Susan Gnora	Saint Michaels
Milton Elmer Hall	Fells Church, Fellsburg
Emma Hobbs	Monongahela

Ignace Hollowiti	Saint Michaels
George Hvizdak	Saint Michaels
Jane Kirkwood	Monongahela
Marcel Kraska	Calvary, Cleveland, Ohio
Andrew Odelga	Saint Dominic
Ida Orr	Monongahela
Thomas Short	Monongahela
Peter Starcovich	Saint Dominic
Perry Stevens	Braddock
Sawka Trubalis	Monongahela
John West	Monongahela

NOTES

Prologue

p. xvii, **"I got out a cloth and went to work."** Berton Roueché, "The Fog," *New Yorker*, September 30, 1950, 298.

p. xviii, **"My God, it just lay there!"** Roueché, "The Fog," 298.

p. xviii, **"You walk in front and show the way."** Roueché, "The Fog," 306.

p. xx, **"Do you think a little smoke is going to bother them?"** Brian Charlton, "Commemorating the 1948 Donora Smog Disaster," *Donora Historical Society*, October 28, 2018.

p. xi, **" . . . suffered another prolonged temperature inversion . . ."** "60 Sudden Deaths Caused by Fog," *Guardian*, December 6, 1962, 1.

1: Donner Takes the Reins

p. 5, **"They included the likes of . . ."** Tom Nicholas and Vasiliki Fouka, "John D. Rockefeller: The Richest Man in the World," *Harvard Business School Faculty and Research*, December 2014, 19.

p. 6, **"There was virtually no tinplate manufactured . . ."** "Tin Plate Industry—Newcastle, PA," Lawrence County Memoirs, 2011, lawrence-countymemoirs.com.

p. 6, **"Donner recalled that . . ."** William H. Donner, *The Autobiography of William Donner* (San Francisco: Cranium Press, 1973), 71.

p. 6, **"He told Donner that . . ."** Donner, *Autobiography*, 71.

p. 7, **"After considering my financial decision . . ."** Donner, *Autobiography*, 70–71.

p. 8, **" . . . River with High Banks . . ."** Arthur Parker, *The Monongahela: River* *of Dreams, River of Sweat* (University Park: Pennsylvania State University Press, 1999), 8.

p. 8, **"... about twenty miles ..."** All distances in this book reflect direct measurements, not distances by roadways. Calculations have been made using Google Maps, either from a specific address or from the approximate center of a town or city.

p. 8, **"Donora was then and remains today ..."** There is, quite coincidentally, a town in Italy named Donoratico, or "little donor," along the Mediterranean coast in Tuscany. Otherwise the name Donora is unique worldwide.

p. 9, **"... later renamed the Union Trust Company."** David Cannadine, *Mellon: An American Life* (New York: Vintage, 2006), 96.

p. 9, **"... monopoly, conspiracy, or even trust."** "Sherman Anti-Trust Act (1890)," Our Documents, National Archives and Record Administration, ourdocuments.gov.

2: Breaking Records

p. 10, **"1223 CARNEGIE BUILDING"** Advertisement, *Pittsburgh Daily Post*, August 7, 1900, 8.

p. 11, **"Much of Donora's coke ..."** Douglas A. Fisher, *Steel Making in America* (New York: United States Steel, 1950), 31.

p. 11, **"... series of nine locks and dams ..."** "Braddock Dam, HAER No. PA-635," Historic American Engineering Record, Northeast Regional Office, National Park Service, 2.

p. 11, **"American Steel & Wire advertised its Ellwood fencing ..."** *Pacific Rural Press* 58, no. 13 (September 23, 1899): 207.

p. 11, **"Donner, wanting to prove his worth to Mellon ..."** "Similar to Charleroi. New Town above Monongahela Will be Big. Outline of Plans," *Daily Republican*, May 31, 1900, 4.

p. 12, **"In many instances there were two or more men camped on the same lot ..."** Roman E. Koehler, "The Way It Was," excerpts of unpublished comments at a dinner at the Donora Lions Club in 1951, courtesy of Donora Historical Society.

p. 12, **"... who had named the town of Monessen."** When Colonel Schoonmaker needed a name for the town he had built and in which William Donner had built a tinplate mill, he asked his friend Henry Clay Frick for ideas. Frick suggested Essen, a large steelmaking city in Germany. That name, however, had already been taken by a tiny locale south of Pittsburgh, now part of Upper Saint Clair. Schoonmaker then decided that Essen be incorporated into the name of the river it lies along, and so was born "Essen on the Monongahela," or Monessen (Donner, *Autobiography*, 60).

p. 12, **"The torrent crushed houses . . ."** David McCullough, *The Johnstown Flood: The Incredible Story behind One of the Most Devastating Disasters America Has Ever Known* (New York: Simon and Schuster, 1968), 146.

p. 12, **"The *Boston Globe* called him . . ."** "Plague Terrors Take the Place of Flood Horrors," *Boston Globe*, June 7, 1889, 4.

p. 12, **"Elmer J. Iiams, an engineer in Donora . . ."** Interview with Rose Marie Iiams, December 18, 2019.

p. 13, **"So he measured the distance that each could raise her legs . . ."** Interview with Rosemarie Iiams, September 16, 2017.

p. 13, **"Pedestrians had to walk . . ."** Sidney Mishkin, *Memories of Donora: Growing Up Jewish in a Western Pennsylvania Steel Town* (unpublished), 23.

3: A Town Blossoms

p. 16, **"There has, probably, never been another instance . . ."** Roman E. Koehler, ed., *Donora Illustrated* (Donora: Donora American, Supplement, 1903), 2.

p. 16, **"It began business July 15, 1901 . . ."** *Banks and Bankers of the Keystone State*, (Pittsburgh: Finance Company in Pittsburgh, 1905), 190.

p. 16, **"He married Mary Magdalena Rugh . . ."** Koehler, *Donora Illustrated*, 1.

p. 17, **"constantly widening circle of Donora's business contingent . . ."** Koehler, *Donora Illustrated*, 9.

p. 17, **"The basement and first floor were 'literally crowded . . .'"** Koehler, *Donora Illustrated*, 12.

p. 17, **" . . . SLOVENSKY POHRABNIK . . ."** "Undertaker" or "mortician" in Slovenian have different spellings, including *pograbnik, podjetnik*, and *pogrebnik*. Why an *h* was used in the sign rather than a *g* or *d*, is unclear. Charles E. Stacey, Brian Charlton, and David Lonich, *Images of America: Donora* (Charleston: Arcadia, 2010), 33.

p. 17, **" . . . building that would perform a quite different function . . ."** "Donora Lumber Company, Inc. in 71st Year of Service to Valley," *Daily Republican*, August 3, 1970, 6.

p. 18, **"The color of the bricks varied . . ."** Emails with Mark Pawelec, September 21, 2020.

p. 18, **"Donora's trolley line was a single spur . . ."** "The Story of Pittsburgh Railways Interurban Division," Charleroi Interurban, http://charleroiinterurban.com.

p. 18, **"What was once the exclusive domain of the 'Idle Classes,' . . ."** Brian Charlton, "Eldora Park Walking Tour," Donora Historical Society and Smog Museum, https://sites.google.com/site/donorahistoricalsociety/.

p. 18, **" . . . doubled as a roller skating rink . . . ,"** Christopher Buckley, "Memories of Eldora Park: Trolley line retreat grew over decades in Carroll Township," *Herald-Standard*, July 20, 2016, heraldstandard.com.

p. 19, **"It was the time of the 'free lunches' . . ."** Ron Paglia, "Donora's Irondale Hotel Was Razed in the Name of Progress," *TribLive*, February 17, 2011, https://archive.triblive.com/. The hotel was razed in 1966 to make room for a branch of the Pittsburgh National Bank.

p. 19, **"Children in the area initially went . . ."** The first school in the area was a log cabin school built in 1790 near the river along the northernmost bend.

p. 19, **"Steppling had been born in . . ."** *75th Diamond Jubilee, 1902–1977: Saint Charles Roman Catholic Church*, Saint Charles Parish, 1977, 8.

p. 21, **"Immigrants from Croatia . . ."** Stacey et al., *Images of America*, 39.

p. 21, **"Many members of those clubs . . ."** *Banks and Bankers of the Keystone State*.

p. 21, **"Those cities tended to develop . . ."** Michael Weber, John Bodnar, and Roger Simon, "Seven Neighborhoods: Stability and Change in Ethnic Community, 1930–1960," *Western Pennsylvania History*, April 1981.

p. 21, **"Although the Spanish did settle near the zinc plant . . ."** Email from Brian Charlton, May 16, 2020.

p. 22, **"Walking in downtown Donora . . ."** Mishkin, *Memories of Donora*.

4: Mavericks Not Allowed

p. 23, **"'Private schools, camps, colleges . . . '"** Johnathan D. Sarna and Jonathan Golden, "The American Jewish Experience in the Twentieth Century: Antisemitism and Assimilation," National Humanities Center, http://nationalhumanitiescenter.org/.

p. 23, **"Resident Sidney Mishkin recalled that Donora stores owned by Jews . . ."** Jewish store owners generally did not use their Jewish-sounding last names to identify their businesses. It was "Bill's Appliance Store" (William "Bill" Burncran), "Henry's Sporting Goods" (Henry Perlet),

"Donora Furniture Company" (Sam and Joe Koenigsberg), "Donora Bowling Alleys" (Benjamin Friedlander).

p. 24, **"Even the great Benjamin Franklin . . ."** Taylor McNeil, "The Long History of Xenophobia in America," *Tufts Now*, September 24, 2020, https://now.tufts.edu/.

p. 24, **"From the time Donora was established . . ."** Douglas O. Linder, "Lynchings: By Year and Race," Famous Trials, https://famous-trials.com/.

p. 24, **"Most of the lynchings occurred . . ."** "Number of Executions by Lynching in the United States by State and Race between 1882 and 1968," *Statista*, statista.com.

p. 24, **" . . . including fourteen in Pennsylvania."** "Lynchings a Part of Pennsylvania's History," *Daily Item*, April 25, 2018, dailyitem.com.

p. 24, **"Donora factories employed many Blacks . . ."** "The Great Migration," History.com, June 28, 2021.

p. 24, **"I never went to a sit-down restaurant . . ."** Phone interview with Reggie Walton, March 12, 2020.

p. 24, **"Blacks couldn't get those jobs."** Phone interview with Helen Jenkins, March 20, 2020.

p. 25, **"I think the kids got along just fine."** Phone interview with Marvin Preston, September 13, 2018.

p. 25, **"We had a skating rink up in Belle Vernon . . ."** Jennifer Bails, "People-Loving Woman Was Perfect to Run Rink," *Trib Live*, January 16, 2006, https://archive.triblive.com/.

5: Building the Mills

p. 27, **"In all, the wire and mills produced . . ."** Clair E. Getty Jr., "Steel, Zinc Workers On Job Turning Out Material for Peace," *Pittsburgh Sun-Telegraph*, October 14, 1947, 1.

p. 27, **"Donner appointed as the first superintendent . . ."** Farrell's brother, James, was also heavily involved in the steel industry and would become president of United States Steel Corporation from 1911 to 1932.

p. 28, **" . . . were owned by American Steel & Wire . . ."** The full name of the company was American Steel & Wire Company, Donora Works, a Subsidiary of United States Steel.

p. 28, **"The team bought eight machines . . ."** Donner, *Autobiography*, 96.

p. 29, "... American Steel & Wire dropped its patent infringement suit ..." Donner, *Autobiography*, 97.

p. 30, "Sheets and strips were used to manufacture ..." Fisher, *Steel Making in America*, 71.

p. 30, "Construction of two blast furnaces ..." Brian Charlton, *Donora Historical Society*, www.sites.google.com/site/donorahistoricalsociety/.

p. 30, "For instance, the floor of Lake Superior ..." Martha S. Carr and Carl E. Dutton, "Iron-Ore Resources of the United States including Alaska and Puerto Rico, 1955," *USGS Publications*, https://doi.org/10.3133/b1082C.

p. 30, "... limestone, a whitish or gray rock ..." Hobart M. King, "Limestone: What Is Limestone and How Is It Used?" Geology.com.

p. 33, "Ammon had lived in the woodlands ..." "Stayed Home Long Time," *Daily Republican*, April 13, 1916, 4.

p. 35, "Jacob eventually became senile ..." Emails with Erin Lavin, April 11, 2020.

p. 35, "Ferdinand had been heir to the Austro-Hungarian Empire ..." "Austria-Hungary," *New World Encyclopedia*, www.newworldencyclopedia.org.

p. 35, "He believed that if it became necessary ..." Mitchell Yockelson, "Pre-War Military Planning (USA)," *1914–1918 Online: International Encyclopedia of the First World War*, October 8, 2014, https://encyclopedia.1914–1918-online.net/.

p. 36, "Older miners might remember calling the ore zinc blende ..." Hobart M. King, "Sphalerite: The Primary Ore of Zinc and a Gemstone with a 'Fire' That Exceeds Diamond," Geology.com.

p. 36, "Persians around the turn of the fourteenth century ..." Marinanne Schönnenbeck and Frank Neumann, "History of Zinc, Its Production and Usage," *Bericht Report*, rheinzink.com.

p. 37, "Melouin said at the time ..." "The Zinc: A Short History of Hot Dip Galvanizing," Berg Banat, www.bergbanat.ro/.

p. 37, "The remaining depressions were sealed ..." Jason Treat, Matt Chwastyk, and Kelsey Nowakowski, "Finding Clotilda," *National Geographic*, May 2019, nationalgeographic.com.

p. 37, "The stamped G.I. is said to have assumed ..." Elizabeth Nix, "Why Are American Soldiers Called GIs?" *History*, www.history.com/news/why-are-american-soldiers-called-gis.

p. 37, "Donora Zinc Works consumed a forty-acre footprint ..." Donald I.

Bleiwas and Carl DiFrancesco, "Historical Zinc Smelting in New Jersey, Pennsylvania, Virginia, West Virginia, and Washington, D.C., with Estimates of Atmospheric Zinc Emissions and Other Materials," Open-File Report 2010-1131, US Department of the Interior and US Geological Survey, 24.

6: Peopling the Mills

p. 39, **"In the early 1900s Rocca Pia . . ."** "Genealogy in Rocca Pia," *Italian Side*, italianside.com.

p. 40, **"Charles Rumford learned to detect that odor . . ."** Charles Rumford Walker, *Steel: The Diary of a Furnace Worker* (Boston: Atlantic Monthly Press), 88–89.

p. 41, **"Living within a single mile of each other . . ."** Amy Roberts, "Here's Where to Find Pittsburgh's Most Historic Mansions (and Tasty Cocktails Nearby)," *Urbanist*, https://pittsburgh.urbanistguide.com/.

p. 42, **"Adella is said to have begun an affair . . ."** "Donner Matrimonial Triangle Is Traced to the Love of the Wife to Society. Divorce Is Granted," *Star and Democrat*, January 18, 1907, 2.

p. 42, **"The hospital physicians have little hope . . ."** "J. Norwood Rodgers Was Probably Fatally Hurt," *Pittsburgh Press*, January 14, 1905, 2.

p. 42, **"Rodgers died six days later . . ."** "J.N. Rodgers Dies of Injuries," *Pittsburgh Daily Post*, January 21, 1905, 6.

p. 43, **". . . Dora's former father-in-law . . ."** William Berlean Rodgers would go on to earn great wealth through investments in coal. He served for many years as president of the Pittsburgh Coal Exchange and also played an integral role in constructing the lock and dam system that ran from Pittsburgh to Cairo, Illinois.

p. 43, **"Both of us being manufacturers of iron . . ."** Donner, *Autobiography*, 122.

p. 43, **"The pair encountered one another several times . . ."** "The Hotel That Saw It All," Shepheard's Hotel, http://shepheard-hotel.com/history.html.

p. 43, **"Donner eventually purchased the plant . . ."** Spencer D. Morgan, *Western New York Steel* (Mount Pleasant, SC: Arcadia, 2014), 9.

p. 44, **"President Herbert Hoover famously labeled . . ."** "The Great Depression," *American Experience*, www.pbs.org/wgbh/american experience/features/dustbowl-great-depression.

p. 44, **"In 1919 the United States produced fully half . . ."** *Impossible Peace: The*

Time between World Wars, episode 1, "The Lap of the Gods, 1919–1921," 2017, 3:37.

p. 44, **"Unemployment stood at just 6.9 percent . . ."** Committee on Economic Security, "Social Security in America: Chapter III, Estimates of Unemployment in the United States," www.ssa.gov.

p. 44, **"In early October 1929 . . ."** "Stock Market Crash of 1929," Federal Reserve History, www.federalreservehistory.org.

p. 44, **"When Wall Street took that tail spin . . ."** Bennett Lowenthal, "The Jumpers of 29," *Washington Post*, October 25, 1987, washingtonpost.com.

p. 45, **"My body should go to science . . ."** Christopher Klein, "1929 Stock Market Crash: Did Panicked Investors Really Jump From Windows?" History.com, March 7, 2019.

p. 45, **"Robert quickly secured a job . . ."** Phone interview with Pat Dugan, August 19, 2019.

7: Wooden Shoes and an Oatmeal Lunch

p. 46, **"He immigrated to the United States in 1910 . . ."** Lesley Kennedy, "Most Immigrants Arriving at Ellis Island in 1907 Were Processed in a Few Hours," History.com.

p. 46, **"He was born and raised in Castrillón . . ."** Phone interview with Rick Lewis (grandson), May 3, 2019.

p. 47, **"Asturias and its neighboring region . . ."** "Mining Historical Heritage," UNESCO World Heritage Center, https://whc.unesco.org/en/tentativelists/5139.

p. 47, **"Rich mineral deposits had been found centuries ago . . ."** "Castile-La Mancha: Heartland of Southern Central Spain," About Spain, https://about-spain.net/tourism/castile-mancha.htm.

p. 47, **"Most of them were skilled mine workers or metallurgists . . ."** "Welcome, Bienvenido, Bienveníu!" Asturian-American Migration Forum, https://www.asturianus.org.

p. 48, **"Manuel and Jiggs found work alongside Frank . . ."** Rich Lewis, "Donora PA 1948 Smog Victim," unpublished.

p. 49, **"The smelter held nine of those furnaces . . ."** Bleiwas and DiFrancesco, "Historical Zinc Smelting," 77.

p. 49, **"Every eight hours a furnaceman . . ."** There is a fascinating video about horizontal retorts called "Zinc Works at Swansea" (YouTube). Shot in the late 1950s, perhaps in 1960, in Swansea, Wales, the film offers a compelling look at how retorts were emptied, cleaned, and replaced. That film and others have been digitized by Periscope Film, a company that archives "endangered non-fiction films." Visit them at www.patreon.com/PeriscopeFilm.

p. 50, **"Another charge would be added . . ."** Bleiwas and DiFrancesco, "Historical Zinc Smelting," 7785; Aspen Junge and Rick Bean, "A Short History of the Zinc Smelting Industry in Kansas," Kansas Department of Health and Environment, December 28, 2006, 4.

p. 50, **"By 1920 the Zinc Works was producing . . ."** Bleiwas and DiFrancesco, "Historical Zinc Smelting," 7777.

p. 50, **" . . . more than any other zinc plant in the world."** Bleiwas and DiFrancesco, "Historical Zinc Smelting," 7774.

p. 50, **"At one point Donora's retorts devoured . . ."** Bleiwas and DiFrancesco, "Historical Zinc Smelting," 77.

p. 50, **" . . . called Broken Hill, some ten thousand miles away."** Bleiwas and DiFrancesco, "Historical Zinc Smelting," 7776. The town of Broken Hill today is perhaps best known for being the main site of filming for the 1994 hit movie *The Adventures of Priscilla, Queen of the Desert*, a frolicking tale about a pair of drag queens (played by Hugo Weaving and Guy Pearce) and a transgender woman (played by Terrance Stamp) traveling and performing across the Australian Outback.

p. 50, **"They each invested seventy pounds . . ."** "Syndicate of Seven," Daytrippa, www.daytrippa.com.au/.

p. 50, **" . . . Broken Hill Proprietary Company, or BHP . . ."** "Broken Hill Mining," Darling River Run, https://discoverbrokenhill.com.au/.

p. 50, **"Fluorine-based minerals can damage . . ."** Bleiwas and DiFrancesco, "Historical Zinc Smelting," 7776.

p. 51, **"No single supplier could provide enough zinc . . ."** Bleiwas and DiFrancesco, "Historical Zinc Smelting," 7777.

p. 51, **"The skin on his face was constantly ruddy . . ."** Lewis, "Donora PA 1948 Smog Victim."

p. 51, **"In the first process, called roasting . . ."** Emails with Art Larvey, March 21, 2020.

p. 51, **"The zinc also bound to oxygen . . ."** The heat generated from those chemical reactions was enough to maintain the temperature inside the roasters without adding additional heat from external sources.

p. 52, **"Like bread on a conveyor toaster . . ."** Donald M. Levy, *Modern Copper Smelting* (London: Charles Griffin, 1912), 60.

p. 52, **"If there was a hell hole in a retort plant . . ."** Emails with Art Larvey, February 2020.

p. 53, **"Even short-term exposure can cause . . ."** "Hazardous Substance Fact Sheet: Sulfur Trioxide," New Jersey Department of Health, https://nj.gov/health/eoh/rtkweb/documents/fs/1767.pdf; "OSH Answers Fact Sheets," Canadian Centre for Occupational Health and Safety, www.ccohs.ca/.

p. 53, **"Attendance was the worst . . ."** Emails with Arthur Larvey, February 2020.

p. 54, **"Furnacemen like Harry Lewis were directly exposed . . ."** Emails with Arthur Larvey, February 8, 2020.

p. 55, **"The exact mechanism . . ."** Keith A. Lafferty and Joe Alcock, "What Is the Pathophysiology of Metal Fume Fever Caused by Smoke Inhalation?" Medscape Nurses, www.medscape.com/.

p. 55, **"In fact, the symptoms of metal fume fever . . ."** M. I. Greenberg and D. Vearrier, "Metal Fume Fever and Polymer Fume Fever," *Clinical Toxicology* 53, no. 4 (May 2015), www.ncbi.nlm.nih.gov/pubmed/25706449.

p. 56, **"Common cures for metal fume fever . . ."** Discussion with Brian Charlton, Donora Historical Society, October 12, 2017.

p. 56, **"Other foods would make him sick . . ."** Lewis, "Donora PA 1948 Smog Victim."

8: Mr. Edison Arrives

p. 57, **" . . . and perhaps go as high as 25,000."** Brian Charlton, "Cement City: Thomas Edison's Experiment with Worker's Housing in Donora," *Western Pennsylvania History* 96, no. 3 (Fall 2013): 37.

p. 57, **" . . . a field of fifty acres . . ."** The area was actually about forty acres. Bleiwas and DiFrancesco, "Historical Zinc Smelting," 74.

p. 57, **" . . . housing and boarding condition in Donora . . ."** "Unable to House Workers at Donora," *Pittsburgh Post-Gazette*, August 2, 1915, 5.

p. 57, **"Some are even traveling daily from Pittsburgh . . ."** "Unable to House Workers at Donora," 5.

p. 58, **" . . . a 'house famine.'"** "Coral St. Houses Will Cost $12,000," *Pittsburgh Press*, June 11, 1916, 33.

p. 58, **"The land between the Pennsylvania Railroad ..."** Donner, *Autobiography*, 93.

p. 58, **"The two important components for growth ..."** "Smart Growth in Small Towns and Rural Communities," US Environmental Protection Agency, www.epa.gov.

p. 59, **"... but was routinely practiced there."** Charlton, "Cement City," 37.

p. 59, **"Thomas Alva Edison was arguably America's greatest innovator ..."** "Thomas Edison: Timeline," National Park Service, www.nps.gov.

p. 59, **"... already invented the phonograph ..."** Natalie Colarossi, "19 Incredible Things You Never Knew Thomas Edison Invented," *Business Insider*, January 31, 2020, businessinsider.com.

p. 59, **"The object of my invention ..."** Thomas A. Edison, "Process of Constructing Concrete Buildings," US Patent 1219,272, filed August 13, 1908, and issued March 13, 1917.

p. 60, **"Edison had, after all, promised ..."** Eric Allen, "Innovative Prototypes for Homes and Why They Didn't Work," *Architectural Digest*, December 15, 2016, architecturaldigest.com.

p. 60, **"First patented in 1824 by Joseph Aspdin ..."** J. L. C. Staff, "Who Invented Cement," *Journal of Light Construction*, May 8, 2019, jlconline. com.

p. 60, **"Aspdin named his invention ..."** Portland cement has become so commonplace in construction that its name has become generic and is no longer capitalized.

9: Walls of Slag

p. 61, **"To all whom it may concern ..."** Edison, "Process of Constructing Concrete Buildings."

p. 61, **"The panels were so heavy ..."** Charlton, "Cement City," 37.

p. 61, **"'Edison's panel forms made the site ..."** Charlton, "Cement City," 39.

p. 62, **"The cement would be mixed ..."** Charlton, "Cement City," 39.

p. 62, **"Slag had been commonly mixed with cement ..."** David F. Salisbury, "Some Smelter Slags Represent a Significant Environmental Hazard," *Stanford University News Service*, December 9, 1998, https://news.stanford.edu/.

p. 62, **"The final ingredients in the concoction ..."** "Role of Gypsum in Cement and Its Effects," *The Constructor*, https://theconstructor.org/.

p. 62, "... absorbent clay used to suspend slag ..." Charlton, "Cement City," 40.

p. 62, "... heavier components of the concrete mixture ..." Matthew Josephson, *Edison: A Biography* (Lexington, MA: Plunkett Lake, 2019).

p. 63, "Edison's experiments had shown ..." Robert Courland and Dennis Smith, *Concrete Planet: The Strange and Fascinating Story of the World's Most Common Man-Made Material* (Buffalo: Prometheus, 2011), 239.

p. 63, "Together, all of that portland cement ..." Charlton, "Cement City," 39.

p. 63, "Brandt was an experienced engineer ..." "The Forbes Field Years," Pirates.com, mlb.com/pirates/.

p. 63, "The boost helped ..." Charlton, "Cement City," 40.

p. 64, "Prairie-style houses could be built so easily ..." Rosemary Thornton, "Do You Have a Sears Kit Home? Tips for Identifying Sears Catalog Houses," Arts and Crafts Society, 2007, www.arts-crafts.com/archive/kithome/rt-searskits.shtml.

p. 64, "Over the 32 years the company offered home kits ..." Thornton, "Do You Have a Sears Kit Home?"

p. 64, "Cement City homes each had a basement and porch." "Cement City Historic District," Living Places, livingplaces.com.

p. 64, "Each house came originally with a lilac bush ..." "Cement City Turns 100 Years Old: Donora Historical Society to Hold Walking Tour of Neighborhood This Month," *Observer-Reporter*, April 5, 2017, https://observer-reporter.com/.

p. 65, "When I was a kid you just didn't walk into Cement City ..." Charlton, "Cement City," 41.

p. 65, "One resident remembers that when Musial visited Donora ..." Charlton, "Cement City," 42.

p. 65, "By comparison a study by the US Census Bureau ..." Na Zhao, "How Long Does It Take to Build a Single-Family Home?" National Association of Home Builders, August 17, 2015, http://eyeonhousing.org/.

p. 66, "Laborers and other lower-level workers ..." Charlton, "Cement City," 40.

p. 66, "Mill officials who rented half of the smallest duplexes ..." Charlton, "Cement City," 40.

p. 66, "In comparison, a New York City apartment in the 1910s ..." "Change Is a Constant in a Century of New York City Real Estate," Miller Samuel, Inc., www.millersamuel.com/files/2012/10/DE100yearsNYC.pdf/.

p. 66, **"Galbreath would go on to build . ."** Warrant Corbett, "John Galbreath," Society for American Baseball Research, https://sabr.org/node/27082.

p. 67, **"Some will ask, 'Is that in Cement City?'"** Charlton, "Cement City," 43.

p. 68, **"Engineers and engineering students traveled to town . . ."** "Hundred Residences Built at Record Pace," *Popular Mechanics*, January 1918, 24–26.

p. 69, **"Perhaps he felt a kinship with the underdog . . ."** "Thomas Edison," Biography.com, April 27, 2017.

p. 69, **"His visits as an adult to the ghettos . . ."** Michael Peterson, "Thomas Edison's Concrete Houses," *American Heritage's Invention and Technology* 11, no. 3 (1996), inventionandtech.com.

10: Transporting Treasures

p. 73, **"Except for a brief depression . . ."** Thomas E. Woods Jr., "The Forgotten Depression of 1920," Mises Institute, November 27, 2009, https://mises.org/.

p. 73, **" . . . unemployment through the decade . . ."** Stanley Lebergott, "Annual Estimates of Unemployment in the United States, 1900–1954," in *The Measurement and Behavior of Unemployment* (Cambridge, MA: National Bureau of Economic Research, 1957), 215, www.nber.org/chapters/c2644.pdf.

p. 73, **"The nation's wealth doubled during the decade."** "The Roaring Twenties History," History.com, February 28, 2020.

p. 73, **"Italian radicals Nicola Sacco and Bartolomeo Vanzetti . . ."** "Sacco and Vanzetti Executed," History.com, July 28, 2019.

p. 74, **"The couple persevered . . ."** Tragedy struck the family again when Billy was killed in a freak accident in 1937. The ten-year-old was on his way to buy some penny candy with a nickel his dad had given him when he tripped on the tracks. He fell just as a switching engine was moving through the area. The wheels cleaved the boy diagonally, from shoulder to hip. The coroner brought the body to the Hart home and placed it on the family's heavy wooden library table. In small towns without a funeral home, people who died were often laid out at home, so the body could be cleansed and the family could stand vigil overnight. Howard and Iva must have felt unimaginable grief as they watched Billy's mutilated body, his knickers still on, placed on the table they had dined on, folded laundry on, and gathered around with friends. And again, Iva was pregnant.

She would deliver her second daughter, Nancy Lee, just five months later. Howard carried a burning guilt inside him until the day he died, believing that Billy's death had been his fault. Howard's last words embodied his anguish: "I should have never given Billy that nickel."

p. 74, **"That section ran along the western shore . . ."** "Pennsylvania Railroads," American Rails, american-rails.com.

p. 74, **"The Pennsy system continued to grow . . ."** "American Railroads," American Rails, american-rails.com.

p. 75, **"And there are locomotives, all kinds of locomotives . . ."** William B. Hard, "Making Steel and Killing Men," *Everybody's Magazine* 17 (November 1907): 586.

p. 75, **"In the earliest years of the factories . . ."** Stacey et al., *Images of America*, 18.

p. 75, **"When no more slag could be dumped there . . ."** "Palmer Park, Donora, Opened in Presence of 15,000 People," *Pittsburgh Post-Gazette*, August 21, 1921, 7.

p. 75, **" . . . park's hugely successful launch."** Ron Paglia, "Original Opening of Palmer Park Was a Big Event," *Tribune-Review*, March 1, 2012, https://archive.triblive.com/.

p. 76, **"Two other races defy explanation . . ."** Paglia, "Original Opening."

p. 77, **"Slag trains consisted of an engine . . ."** "Steel Train Dumping Liquid Molten Slag," Train Fanatics, https://trainfanatics.com/steel-train-dumping-liquid-molten-slag.

p. 77, **"The bell would crash apart on the hill . . ."** For a video of this remarkable process, see "Dumping Slag at Bethlehem Steel in 1994," YouTube.

p. 78, **"Steel mills in Braddock, Duquesne, and Homestead . . ."** Roy Kohler, "Waste Slag May Be Hero in Road Building," *Pittsburgh Post-Gazette*, January 22, 1950, 42.

p. 78, **"If the Brooklyn Bridge were made purely of slag . . ."** The Brooklyn Bridge weighs about 14,680 tons. Not long after the bridge opened in 1884, P. T. Barnum led twenty-three of his elephants, including the famous Jumbo, in a massive parade across the bridge to prove its stability.

p. 78, **"Still another recalled that . . ."** "Brown's Dump: The West Mifflin Mountain of Slag," Brookline Connection, brooklineconnection.com.

11: Ol Little Town of Webster

p. 79, **"Yards were separated by white picket fences . . ."** Scott Beveridge,

"The Warning Signs Were There," *Travel with a Beveridge* (blog), August 5, 2007, https://scottbeveridge.blogspot.com/.

p. 80, **"The men fought to maintain peace . . ."** "Three Senatorial Giants: Clay, Calhoun and Webster," U.S. History, www.ushistory.org.

p. 80, **"The original Beazells . . ."** The family has never quite settled on a single spelling of their common surname. John Hepple, a descendant of the original Beazells and member of the Rostraver Historical Society, said, "In the Fells Cemetery there are four different spellings of the name. In most cases even brothers spelled it differently." Email from John Hepple, June 24, 2020.

p. 80, **" . . . from Basel, an ancient city along the Rhine River . . ."** Although historical documents for the Beazell family commonly list "Basel, Germany" as the family's original hometown, Basel itself was never part of Germany, a nation-state founded in 1871. More likely, the "Germany" in the documents originated from Basel residents feeling a greater affinity for the Germanic culture. In the eighteenth century, when Matthew Basel emigrated, Basel was part of the Holy Roman Empire in an area now known as Switzerland.

p. 81, **"The Webster Roller Flour Mill . . ."** Ad, *Monongahela Valley Republican*, April 3, 1902, 6.

p. 81, **"Apparently Byers was 'an eye specialist . . . '"** "Webster Notes," *Valley Independent*, October 20, 1906, 4.

p. 82, **"Gilmore died in 1884 . . ."** John M. Gresham, *Biographical and Historical Cyclopedia of Westmoreland County, Pennsylvania* (Philadelphia: John Gresham, 1890), 588.

p. 82, **"That classy hotel touted a lavish restaurant . . ."** Beveridge, "Warning Signs."

p. 82, **"Jacob Tomer and his wife, Harriet . . ."** Email from Scott Beveridge, June 16, 2020.

p. 83, **"John Vogel, who ran the Union Hotel . . ."** Ad, *Monongahela Valley Republican*, April 3, 1902, 6.

p. 83, **"He enlisted in the volunteer army in 1861 . . ."** Gresham, *Westmoreland County*, 643–44.

p. 83, **" . . . in a series of battles . . ."** William Miller, "The Seven Days Battles," *American Battlefield Trust*, www.battlefields.org.

p. 83, **" . . . the celebrated veteran built the Union Hotel . . ."** "Hands across the River," *Pittsburgh Press*, December 5, 1908, 2.

p. 83, **"The *Daily Independent* proclaimed the renovations . . ."** "Hotel Extensively Remodeled," *Daily Independent*, June 12, 1907, 4.

p. 84, **"Anyone else with enough money . . ."** Beveridge, "Warning Signs."

p. 84, **"Proponents of the plan circulated a petition . . ."** "Big Celebration Is to Mark Opening of the New Donora-Webster Bridge," *Pittsburgh Press*, November 29, 1908, 16.

p. 84, **"Judge Lucien W. Doty rejected the proposal . . ."** "Big Celebration," *Pittsburgh Press*, 16.

p. 85, **"The Donora-Webster Free Bridge . . ."** "Donora-Webster Bridge," Historic Bridges, https://historicbridges.org/.

p. 85, **"The bridge provided ready access . . ."** The bridge carried vehicles and pedestrians back and forth throughout Donora's history until it was demolished in 2015.

p. 85, **"The *Donora American* reported that . . ."** Scott Beveridge, "Old Bridge Having a Birthday," https://scottbeveridge.blogspot.com/2008/11/old-bridge-having-birthday.html.

p. 85, **"The wedding took place at the midpoint . . ."** "Man Jumped from the Donora-Webster Bridge on Saturday," *Daily Republican*, December 7, 1908, 1.

p. 85, **"He seemed pleased with his experience . . ."** "Man Jumped," 1.

p. 86, **"A smoker was a very exclusive gathering . . ."** Email from Brian Charlton, June 16, 2020.

p. 86, **"The pouch contained a swanky El Solano . . ."** Caswell A. Mayo, ed., *American Druggist and Pharmaceutical Record* (New York: American Druggist, 1911), 57. One of the smoking kits can be viewed at the Donora Historical Society.

12: Zinc in the Wind

p. 87, **"At the height of the smelter's production . . ."** Bleiwas and DiFrancesco, "Historical Zinc Smelting," 87.

p. 88, **"Beveridge recalled that . . ."** Scott Beveridge, "Dying for Fresh Air," *Travel with a Beveridge* (blog), September 2, 2007, https://scottbeveridge.blogspot.com/.

p. 90, **"This Plant is directly across the railroad tracks . . ."** Gilmore Family Papers, courtesy of Clifford Gilmore, October 7, 2021.

p. 91, **"Beveridge testified to the toughness . . ."** Beveridge, "Warning Signs."

13: It Takes a Killing

p. 93, **"Born in Italy in 1887, Giura immigrated . . ."** Certificate of Death: Angelo Guira, Commonwealth of Pennsylvania, Bureau of Vital Statistics, Allegheny County, File 6091, Registered Number 611; Informant: Carmelo Giura, Pittsburgh, PA.

p. 93, **"When he came of age the young man . . ."** Crystal Eastman, *Work-Accidents and the Law* (Chicago: Russell Sage Foundation, 1910), 56.

p. 94, **"His supervisor had considered him . . ."** Eastman, *Work-Accidents,* 56.

p. 94, **"At ten minutes before eight on the evening of January 29, 1907 . . ."** Certificate of Death: Angelo Guira.

p. 94, **"Yet it required the killing of Angelo . . ."** Eastman, *Work-Accidents,* 57.

p. 94, **"The report counted 1,339 injuries . . ."** How many injuries and deaths occurred in Donora is unknown, as are those in any of the other blast furnaces studied. The report described incidents in the aggregate, without mentioning locations.

p. 94, **"Injuries from hand labor . . ."** Frederick H. Willcox, *Occupational Hazards at Blast-Furnace Plants and Accident Prevention Based on Records of Accidents at Blast Furnaces in Pennsylvania in 1915* (Washington, DC: US Government Printing Office, 1917), 116–17.

p. 95, **"Laborer was run over and disemboweled . . ."** Willcox, *Occupational Hazards,* 20–22.

p. 95, **"Nearly twice as many men that year were killed . . ."** Eastman, *Work-Accidents,* 57.

p. 95, **"With so many workers speaking so many different languages . . ."** Willcox, *Occupational Hazards,* 10.

p. 96, **"Frequently fifty percent can not speak English . . ."** Willcox, *Occupational Hazards,* 10.

p. 96, **"Combining the obstacle of not speaking English . . ."** Willcox, *Occupational Hazards,* 10.

p. 97, **"One safety sign at the steel mill in Donora . . ."** Bruce Dreisbach, "A Finished Job," Bruce Dreisbach–American Steel & Wire Company Photographs, University of Pittsburgh, 1915–17, Identifier 200102.110.DR.

p. 97, **"In large printing were the words . . ."** Bruce Dreisbach, "Mill Safety Bulletin Board," Bruce Dreisbach–American Steel & Wire Company Photographs, University of Pittsburgh, 1915–17, Identifier 200102.022.DR.

p. 97, **"It is useless to expect a Slovenian . . ."** William B. Hard, "Making Steel and Killing Men," *Everybody's Magazine* 17 (November 1907): 586.

p. 98, **"The reports were sponsored by the Russell Sage Foundation . . ."** Margaret Olivia Sage, "Letter of the Gift," Russell Sage Foundation, April 19, 1907, www.russellsage.org.

p. 98, **"For instance, about women working in the metal industries . . ."** Elizabeth Beardsley Butler, *Women and the Trades: Pittsburgh, 1907–1908* (Pittsburgh: University of Pittsburgh Press, 1984), 228.

p. 99, **"Freda Kirchwey, longtime editor of *The Nation* . . ."** "Crystal Eastman Keeps Her Name, Doesn't Do Housework," *New England Historical Society*, www.newenglandhistoricalsociety.com/crystal-eastman-keeps-name.

p. 99, **"She graduated from Vassar College in 1903 . . ."** "Crystal Eastman: American Lawyer, Writer, Activist," *Encyclopaedia Britannica*, britannica.com.

p. 99, **"That law became the model for similar laws . . ."** Susan N. Herman, "Crystal Eastman, the ACLU's Underappreciated Founding Mother," ACLU, www.aclu.org/issues/free-speech/crystal-eastman-aclus-underappreciated-founding-mother.

p. 99, **" . . . American Civil Liberties Union . . ."** "Crystal Eastman," National Women's Hall of Fame, www.womenofthehall.org/inductee/crystal-eastman.

p. 99, **"Congress shall have power to enforce this article . . ."** "Frequently Asked Questions," Equal Rights Amendment, www.equalrightsamendment.org.

p. 99, **"The Mott amendment was proposed in every congressional session . . ."** Erin Blakemore, "Why the Fight Over the Equal Rights Amendment Has Lasted Nearly a Century," History.com.

p. 99, **"Eastman succumbed at forty-seven . . ."** Amy Aronson, *Crystal Eastman: A Revolutionary Life* (New York: Oxford University Press, 2020), 2. Eastman was in the early stages of scarlet fever when her older brother, Morgan, perished from the same infection.

p. 100, **"Nearly 40 percent of all fatalities at steel mills . . ."** Eastman, *Work-Accidents*, 72.

p. 102, **"To reduce train accidents . . ."** Eastman, *Work-Accidents*, 58–59.

p. 102, **"Pittsburgh's business and professional elites . . ."** Meryl Nadel, "The Pittsburgh Survey of 1907–1908: Divergent Paths to Change," *Social Service Review* (December 2019): 679.

p. 102, **"As the survey's leader Kellogg put it . . ."** Nadel, "Pittsburgh Survey," 691.

p. 102, **"Nadel wrote that 'Kellogg and the other investigators . . . '"** Nadel, "Pittsburgh Survey," 696.

p. 103, **"Bookstores refused to sell the reports in book form . . ."** Nadel, "Pittsburgh Survey," 697.

14: The Persistent Legend of Young Andrew Posey

p. 104, **"With no war now to fight, Posey was demobilized . . ."** Most tellings of the Posey accident have Andrew returning "home from the war," which may sound heroic but doesn't comport with the facts. According to Andrew's military records, he never left Pittsburgh, at least not while on duty.

p. 105, **"The Posey article read, in its entirety . . ."** "Hot Metal Shower Kills Youth," *Pittsburgh Post-Gazette*, January 10, 1920, 7.

p. 105, **"An inquest was apparently held . . . ,"** "Mill Worker Victim of Accidental Burns," *Donora American*, January 9, 1920, 1.

p. 105, **" . . . but the report has not survived."** Phone call with Washington County Coroner's Office, January 4, 2022. No coroner records prior to 1979 exist.

p. 105, **"According to a 2006 letter . . ."** James McKenzie, "Man of Steel," *Notre Dame Magazine*, Winter 2006–2007, https://magazine.nd.edu/.

p. 107, **"Still another story has Posey signaling the furnaceman . . ."** "Letter to the Editor" in response to Chris Buckley, "Conversation with Donora Official Inspires Cal U Kids," *Tribune-Review*, April 29, 2011. No published copy available.

p. 108, **"He was twenty-two . . ."** "Understanding the Teen Brain," University of Rochester Medical Center, www.urmc.rochester.edu/.

p. 108, **"The interaction of those psychological constructs . . ."** Email from Alin I. Coman, PhD, May 21, 2020.

p. 108, **"People have different positions in their social networks . . ."** Email from Alin I. Coman, PhD, May 21, 2020.

p. 109, **"'As the table was raised,' said Eastman . . ."** Eastman, *Work-Accidents*, 54.

p. 110, **"The company, with a sense of the proprieties . . . "** Walker, *Steel*, 28.

p. 110, **"Lynne Page Snyder, whose 1963 doctoral dissertation . . ."** Lynne Page Snyder, "The Death-Dealing Smog over Donora, Pennsylvania: Industrial Air Pollution, Public Health, and Federal Policy, 1915–1963" (PhD diss., University of Pennsylvania, 1994), 20.

p. 111, "**Eastman pulled no punches in her book . . .**" Eastman, *Work-Accidents*, 72–73.

p. 111, "**The engineer found nothing but . . .**" Scott Beveridge, "Donora 'Man of Steel' Grave Was a Hoax," *Travel with a Beveridge* (blog), http://scottbeveridge.blogspot.com/.

15: Death on the River Meuse

p. 113, "**The entire fountain is surrounded . . .**" "Bassinia Fountain," Ville de Huy: Tourisme, www.huy.be/.

p. 114, "**The façade sculpted on the gate depicts . . .**" "Portal of the Virgin, Says Bethlehem," Ville de Huy: Tourisme, www.huy.be/.

p. 114, "**Animal carcasses lay sprawled about . . .**" Dale Brumfield, "Belgium's Fog of Death," *Medium*, October 22, 2018, https://medium.com/.

p. 114, "**The deaths along the Meuse River shocked not only Belgium . . .**" "Deadly Fog in Belgium. Poison Gas or Factory Fumes?" *Guardian*, December 6, 1930, 11.

p. 114, "**Another British paper, *The Observer*, not only noted the tragedy . . .**" "60 Fog Deaths in Belgium," *Observer*, December 7, 1930, 17.

p. 114, "**The *Daily Republican*, Monongahela's main newspaper . . .**" "Death-Bringing Fog Lifting in Belgium: Dampness Alone Believed to Have Caused or Hastened 42 Deaths in Meuse Valley," *Daily Republican*, December 6, 1930, 1.

p. 115, "**The *Pittsburgh Post-Gazette* the same day . . .**" "Mystery Poison Gas Kills 64 in Belgium: Meuse Valley Inhabitants in Panic as Thick Fog Spreads Death," *Pittsburgh Post-Gazette*, December 6, 1930, 1.

p. 115, "**Theories abounded, including that the deaths were caused . . .**" Peter Brimblecombe, ed., *Air Pollution Episodes* (Singapore: World Scientific, 2017), 30.

p. 115, "**They have been having floods in that district lately . . .**" "Likens Epidemic to 'Black Death' of Middle Ages," *Oil City Derrick*, December 6, 1930, 1.

p. 115, "**The tortuous Meuse flows through Belgium . . .**" "Belgium: Geography," Global Security, www.globalsecurity.org.

p. 116, "**Engis was home to the Métallurgique de Prayon . . .**" The company continues today as Prayon, a major phosphate manufacturer.

p. 116, "**That section of the river valley also contained . . .**" Alexis Zimmer and

Benoit Nemery, "Meuse Valley (1930): Just Fog or Industrial Pollution?" in Brimblecombe, *Air Pollution Episodes*, 31.

p. 116, **"Brussels also held a centenary parade . . ."** "Belgium's Centenary," *New York Times*, July 6, 1930, 37.

p. 116, **" . . . before taking a slow left turn into Place Royale" . . .** Email from Sevgi Ertoy, visit.brussels Tourism Information Desk, March 31, 2020.

p. 116, **"The parade featured columns of marchers . . ."** "'Gallant Little Belgium,' 1930," *British Pathé*, April 13, 2014, youtube.co.

p. 117, **"As the air right above the ground is cooled down . . . "** Phone interview with Neil Donahue, January 5, 2018.

p. 117, **"Ground-level air remained cool . . ."** Benoit Nemery, Peter H. M. Hoet, and Abderrahim Nemmar, "The Meuse Valley Fog of 1930: An Air Pollution Disaster," *Lancet*, March 3, 2001, 704.

p. 118, **"The fog of December 1930 . . ."** Nemery et al., "Meuse Valley Fog," 705.

p. 118, **"The fog reached its thickest on Wednesday . . ."** Nemery et al., "Meuse Valley Fog," 704.

p. 118, **"Hour after hour the doctors saw patient after patient . . ."** "Germans Introduce Poison Gas," History.com.

p. 119, **"Besides the dead, there are hundreds suffering . . . "** Brumfield, "Belgium's Fog of Death."

p. 119, **"The mayor spoke later . . ."** "Belgians in a Panic: A Mist of Death: Heavy Death Roll," *Townsville Daily Bulletin*, December 8, 1930, 6, https://trove.nla.gov.au/newspaper/article/61040719.

p. 119, **"By comparison, Engis . . ."** Zimmer and Nemery, "Meuse Valley," 33.

p. 119, **"Lacombe and Timbal released their surely comprehensive report . . ."** Zimmer and Nemery, "Meuse Valley," 30.

p. 119, **"A noted French physiologist, Jules Amar . . ."** Brimblecombe, *Air Pollution Episodes*, 31.

p. 120, **"Pierre Nolf, the king's physician . . ."** Zimmer and Nemery, "Meuse Valley," 31.

p. 120, **"The Belgian government agreed and formed two committees . . ."** Zimmer and Nemery, "Meuse Valley," 32.

p. 120, **"After testing the concentrations of the compounds . . ."** Nemery et al., "Meuse Valley Fog," 706.

p. 120, **"Firket blamed only . . ."** Nemery et al., "Meuse Valley Fog," 706.

p. 121, **"The statue depicts a young woman . . ."** Nemery et al., "Meuse Valley Fog."

16: Decisions, Decisions

p. 122, **"The bureau told me to buy my hay . . ."** "Smoke Ruin Cited by Farmer," *Pittsburgh Sun-Telegraph*, November 10, 1936, 23, www.newspapers. com/image/524059089.

p. 123, **"A total of 100,362 individual ovens . . ."** "Beehive Coke Oven to Fire Up One Last Time," *New York Times*, September 12, 1982, 73, nytimes. com.

p. 124, **"Ovens in the mill were also able to trap much of the gas . . ."** Snyder, "Death-Dealing Smog," 93.

p. 124, **"'This unconventional claim,' wrote Snyder . . ."** Snyder, "Death-Dealing Smog," 96.

p. 124, **"He wanted to show that each plaintiff's . . ."** Snyder, "Death-Dealing Smog," 97.

p. 125, **"Judge V. B. Woolley, writing for the appellate court . . ."** *Gliwa v. United States Steel Corporation*, 58 F.2d 920, 921 (3d Cir. 1932), https:// casetext.com/.

p. 126, **"Even the term 'smog'. . ."** Murray Whyte, "Dr. Des Voeux and the Invention of Smog," *The Star*, June 10, 2007, thestar.com. A British scientist and anti-pollution activist named Henry Antoine Des Voeux coined the term in 1909, after a "smokey fog" that hung over Glasgow, Scotland, left more than a thousand people dead. Des Voeux contracted "smokey fog" into "smog," the term still used today.

p. 127, **"The company even built an agricultural testing station . . ."** Snyder, "Death-Dealing Smog," 101.

p. 127, **"Judge Woolley even praised Pinkasiewicz . . ."** *Gliwa v. US Steel*.

p. 128, **"Mamie had been a healthy young mother . . ."** *Burkhardt et ux. v. American Steel & Wire Company, Appellant*, Superior Court of Pennsylvania, Vol. 74 (Philadelphia: Geo. T. Bisel, 1921), 440.

p. 129, **"He wrote that Mamie Burkhardt . . ."** *Burkhardt v. American Steel & Wire*, 443.

17: The Days Before

p. 133, **"Germans convicted of war crimes . . ."** Sydney Gruson, "Palestine Accord Foreseen in Israel," *New York Times*, October 15, 1948, 2.

p. 133, **"Casey Stengel took the helm . . ."** "Casey Stengel Quotes," *Baseball Almanac*, www.baseball-almanac.com/quotes/quosteng.shtml.

p. 134, **"The Roper polling company . . ."** "Looking Back at the Truman Beats Dewey Upset," National Constitution Center, November 3, 2017, https://constitutioncenter.org/blog/behind-the-biggest-upset-in-presidential-history-truman-beats-dewey.

p. 134, **"Truman had spoken to a huge crowd . . ."** Associated Press, "Truman, Dewey Reach Stretch," *Poughkeepsie Journal*, October 26, 1948, 1.

p. 134, **"Eternal vigilance is still the price of liberty . . ."** "Address in the Chicago Stadium," American Presidency Project, www.presidency.ucsb.edu/.

p. 134, **"It is a campaign to strengthen . . ."** Robert E. Hartley, *Battleground 1948: Truman, Stevenson, Douglas, and the Most Surprising Election in Illinois History* (Carbondale: Southern Illinois University Press, 2013), 192.

p. 134, **"Twelve years later, in 1930 . . ."** Alonzo L. Hamby, "Harry S. Truman: Life before Presidency," University of Virginia Miller Center, https://millercenter.org/.

p. 135, **"Finishing fourth in her high school class . . ."** "194 Seniors Receive Diplomas Tomorrow," *Charleroi Mail*, May 25, 1945, 1 and 6.

18: The First Days

p. 137, **"During the period of the poison fog . . ."** Bill Davidson, "Our Poisoned Air," *Collier's*, October 23, 1948, 68.

p. 138, **"Donnie Dingbat was a cartoon bird . . ."** Jim Davidson, "The Face Is Familiar, But . . . ," *Pittsburgh Press*, July 20, 1986, 194.

p. 138, **" . . . Roosevelt had instituted restrictions . . ."** "Records of the office of Censorship: 1934–45," US National Archives, www.archives.gov.

p. 138, **"Twenty minutes later Donnie Dingbat was born."** Joe Bennett, "32 Years Under the Weather," *Pittsburgh Press*, March 10, 1974, 256.

p. 138, **"The relative humidity was 99 percent . . . "** "Fog Grounds 'Witch' Dingbat," *Pittsburgh Press*, October 26, 1948, 1. Donnie Dingbat remained a front-page staple until his "retirement" in 1979.

p. 139, **"They tend to flow away . . ."** "The Highs and Lows of Air Pressure," University Corporation for Atmospheric Research Center for Science Education, https://scied.ucar.edu/.

p. 139, **"Anyone reading that day's . . ."** "Lee Sala Wins 44th Bout," *Daily Republican*, October 26, 1948, 2.

p. 139, **" . . . the body of legendary slugger . . ."** "Body of Babe Ruth Finally Laid to Rest," *Daily Republican*, October 26, 1948, 2.

p. 139, " . . . two 'thugs' had stolen $150 . . ." *Daily Republican,* October 26, 1948, 1.

p. 140, "**Chambon owned a moving and storage . . .**" Roueché, "The Fog," 307.

p. 140, "**She loved the cheesecake there . . .** " Email from Annie Schempp, July 14, 2020.

p. 140, "**The Monongahela High School football team . . .**" "Dragon Weight Edge Worries MHS Coaches," *Daily Republican,* October 27, 1948, 2.

p. 141, "**Such a weight advantage . . .** " "Dragon Weight Edge Worries MHS Coaches," 2.

p. 142, "**Fred J. Bliss, on behalf of area residents . . .**" *Kowall v. US Steel,* Case No. 2017-3355, filed July 7, 2017.

p. 142, "**Although Bliss, not surprisingly . . .**" Susan Dunlap, "A Dangerous Job That Gave Life to a Town: A Look Back at the Anaconda Smelter," *Montana Standard,* January 22, 2009.

p. 143, "**Musial finished first in voting that year . . .**" "Stan Musial," Baseball Reference, www.baseball-reference.com/players/m/musiast01.shtml.

p. 143, "**The two Griffeys . . .**" Charles F. Faber, "Ken Griffey, Sr.," Society for American Baseball Research, https://sabr.org/.

p. 143, "**The team had played . . .**" Emails with Amy Scheuneman, executive director, Western Pennsylvania Interscholastic Athletic League, July 10, 2020.

19: Friday

p. 146, "**They had been having trouble breathing . . .**" Bill Davidson, "Donora: The Case of the Poisoned Air," *Collier's,* October 22, 1949, 31.

p. 147, "**He became a charter member . . .**" "Donora Spanish Club Honors M. M. Neale," *Daily Republican,* January 26, 1950, 1.

p. 148, "**Neale also played leading roles . . .**" Program pamphlet of a dinner given by the Distinguished Citizens Club of Donora honoring M. M. Neale, probably 1952 or 1953.

p. 148, "**Neale decided to conduct . . .**" Benjamin Ross and Steven Amter, "Donora's Strangler Smog," in Benjamin Ross and Steven Amter, *The Polluters: The Making of Our Chemically Altered Environment* (New York: Oxford University Press, 2010), 99.

p. 148, "**Without a charge to keep the furnace going . . .**" Email from Art Larvey, July 15, 202.

p. 148, "**Koehler and Roth arrived . . .**" "Mon Valley Medicine: Area Doctors and Hospitals—1938," http://freepages.rootsweb.com/~pamonval/genealogy/files/doctors1938.html.

p. 150, **"The 'young assistant' . . ."** "Physician's Satchel Stolen at Donora," *Daily Republican*, May 1, 1951, 1.

p. 150, **"Suddenly he saw . . ."** Davidson, "Donora," 30.

20: Heroes and Villains

p. 154, **"The Rongauses eventually settled . . ."** Emails from Nancy Rongaus Cherney, October 18, 2020.

p. 155, **"Producing weapons of war . . ."** Logan Nye, "This Top Secret Mission Kept the Nazis from Getting Amsterdam's Diamonds," *We Are the Mighty*, February 19, 2016, wearethemighty.com.

p. 156, **"Nurses from around the hospital . . ."** Emails from Nancy Rongaus Cherney, October 2020.

p. 157, **" . . . refused in June 1939 to allow 937 passengers . . ."** Erin Blakemore, "A Ship of Jewish Refugees Was Refused US Landing in 1939. This Was Their Fate," History.com, June 4, 2019.

p. 157, **"So strong was antisemitic sentiment . . ."** Justin Ewers, "FDR Pushed for the Rescue of Jewish Refugees, Newly Revealed Documents Show," *HistoryNet*, September 2009, historynet.com.

p. 157, **"Then, with impeccable timing . . ."** Phone interview with Jerry Harris, September 4, 2020.

p. 159, **"Can't shove it back in . . ."** Email from Jerry Harris, September 12, 2021.

p. 159, **"Both physicians treated some 'pitiful cases,' . . ."** Roueché, "The Fog," 301.

p. 160, **"Roth ignored them all . . ."** Roueché, "The Fog," 301–2.

p. 160, **"Des Voeux was widely commended . . ."** Whyte, "Dr. Des Voeux and the Invention of Smog."

p. 160, **" . . . he and his family had been forced to leave Donora . . ."** Notes from Devra Davis's papers at the University of Pittsburgh.

p. 161, **"It is a sharp curve . . ."** John R. Bell, "Four Boats in Collision at Webster," *Pittsburgh Post-Gazette*, February 7, 1941, 13.

p. 161, **"Notations on those individuals included . . ."** Notes from Devra Davis's papers at the University of Pittsburgh.

21: Halloween Parade

p. 165, **"Of the hundreds of onlookers . . ."** Roueché, "The Fog," 303.

p. 165, **"Donora's mostly volunteer department . . ."** "About," Donora Volunteer Fire Company, https://donorafirecompany.weebly.com/.

p. 165, **"It had us all coughing."** Roueché, "The Fog," 303.

p. 167, **"I would give one of the patients a treatment . . ."** Snyder, "Death-Dealing Smog," 24.

22: The Town Reacts

p. 169, **"But what the hell . . ."** Roueché, "The Fog," 301–4.

p. 169, **"Panicked, she 'dashed madly . . .'"** "Terribly Burned When Clothing Caught Fire," *Daily Republican*, May 4, 1920, 1.

p. 169, **"She died eighteen years later . . ."** "Newspaper Woman, Former Donoran, Ends Life in N. Y.," *Daily Republican*, May 10, 1938, 1.

p. 169, **"On May 1, 1922 . . ."** "Donora Yard Foreman Probably Fatally Injured," *Daily Republican*, May 1, 1922, 4.

p. 170, **"It eased my throat."** Roueché, "The Fog," 304.

p. 171, **"She was at a party . . ."** Gladys Schempp diary, courtesy of Annie Schempp.

p. 171, **"According to his daughter . . ."** Interview with Annie Schempp, January 6, 2018.

23: Death Begins Its Assault

p. 174, **"Frankly, I don't know how . . ."** Roueché, "The Fog," 305.

p. 175, **"Ceh was a leader in the town's . . ."** James A. Magill, "Notice of Incorporation," *Charleroi Mail*, May 5, 1915, 3.

p. 175, **"His breathing worsened . . ."** H. H. Schrenk, Harry Heimann, George D. Clayton, and W. M. Gafafer, *Air Pollution in Donora, Pa: Epidemiology of the Unusual Smog Episode of October 1948: Preliminary Report* (Washington, DC: US Public Health Service, 1949), 51.

p. 176, **"A longtime member of the Donora School Board . . ."** "Donora Church Club Holds Annual Meeting," *Charleroi Mail*, January 24, 1944, 8.

p. 178, **"He moved to Donora . . ."** Schrenk et al., *Air Pollution in Donora*, 50.

24: "Oh, Helen, My Dad Just Died!"

p. 180, **"He had barely slept . . ."** Davidson, "Donora," 31.

p. 181, **"The dyspnea didn't last long . . ."** Schrenk et al., *Air Pollution in Donora*, 47.

p. 183, **"But she never did."** Phone interview with Edie Jericho, September 8 and 9, 2020.

p. 183, **"That stuff is coming over here . . ."** Snyder, "Death-Dealing Smog," 25.

p. 185, **"So far, however, we've been lucky."** "Smog Causes Many Deaths," *News and Observer*, October 31, 1948, 6.

p. 186, **"Rongaus, for instance, was about . . ."** William Rongaus, "Doctor Tells of Smog Tragedy," *News-Herald*, November 1, 1948, 1.

p. 186, **"Besides Bill Schempp . . ."** "Son of Former Local Couple Weds Bentleyville Girl at Ceremony at Donora Church," *Daily Republican*, June 30, 1953, 3.

p. 186, **"At six feet four . . ."** "Donora H. S. Grid Squad Begins Work," *Daily Republican*, September 4, 1930, 5.

p. 186, **"Some of them didn't give a damn . . . "** Christopher Bryson, *The Fluoride Deception* (New York: Seven Stories, 2004), 117.

p. 186, **"According to her yearbook . . ."** *The Dragon, 1945* (Donora: Donora High School Seniors, 1945), 27.

p. 186, **"Everybody is dying!"** "In the Air," Documentary Works, www.thedocumentaryworks.org.

p. 187, **"The Physicians Exchange started . . ."** "Physicians Open Exchange," *Daily Republican*, December 5, 1938, 1.

p. 187, **"She operated out of the Ostrander family home . . ."** Phone interview with Demitri Petro, MD, September 24, 2020.

p. 187, **"Davis's blanket-over-the-head setup . . ."** Email with Gary White, RRT, MS, March 3, 2019. White was the director of the respiratory care program at Spokane Community College.

p. 187, **"You had to get right up to the door . . . "** David Templeton, "Donora's Doomsday Message," *Pittsburgh Post-Gazette*, October 29, 1998, F-2.

p. 188, **"It almost broke my heart to leave."** Templeton, "Donora's Doomsday Message," F-2.

p. 188, **"I was hardly even coughing much."** Roueché, "The Fog," 307.

p. 189, **"The corps had grown by 1948 . . ."** "Our History," Commissioned Corps of the US Public Health Service, www.usphs.gov.

p. 189, **"The USPHS team finally arrived . . ."** Davidson, "Donora," 60.

25: Game Day

p. 190, **"The game was set to start . . ."** "Cats Will Be after First Win over Dragons Since '42," *Daily Republican*, October 29, 1948, 2.

p. 190, **"The Wildcats were underdogs . . ."** "Valley Scribes Pick Dragons to Whip Cats," *Daily Republican*, October 29, 1948, 2.

p. 190, **"The last time the Wildcats . ."** Kenny Lusk, "Cats Score in Last Half Minute, Beat Dragons, 12–6," *Daily Republican*, November 9, 1942, 2.

p. 190, **"I was worried about the hot dogs . . ."** Templeton, "Donora's Doomsday Message," F-2.

p. 190, **"She needn't have concerned herself . . ."** "Valley Scribes Pick Dragons," 2.

p. 191, **"The only difference between that fog . . ."** Phone interview with Ken Barbao, October 26, 2020.

p. 191, **"Lignelli had graduated from . . ."** Dan Majors, "John 'Chummy' Lignelli: Longtime Donora Mayor Also Lifted People Up," *Pittsburgh Post-Gazette*, December 4, 2016, post-gazette.com.

p. 191, **"But we stayed and watched."** Templeton, "Donora's Doomsday Message," F-2.

p. 191, **"For example, the *Daily Republican's* sportswriter . . ."** Bryson, *Fluoride Deception*, 117.

p. 191, **" . . . described the Monongahela touchdowns . . ."** "Charles Legeza Leads Monongahela to Easy Win Over Donorans," *Daily Republican*, November 1, 1948, 2.

p. 192, **"Stanley was too late; his father was dead."** Devra Lee Davis, "The Heavy Air of Donora, Pa.," *Chronicle of Higher Education*, October 25, 2002, chronicle.com.

p. 192, **"The public-address system . . ."** Davidson, "Donora," 60.

p. 192, **"Sawa played nearly the entire game . . ."** "'Cats Explode Donora Myth, 27–7," *Daily Republican*, September 26, 1948, 7.

p. 192, **"I'm running up these streets . . ."** Phone call with James Sawa, February 13, 2020.

p. 193, **"Hospice professionals term that . . ."** Michael Nahm, Bruce Greyson, Emily Williams Kelly, and Erlendur Haraldsson, "Terminal Lucidity: A Review and a Case Collection," *Archives of Gerontology and Geriatrics* 55, no. 1 (July–August 2012): 138.

p. 193, **"Family members often view this as a miracle . . ."** Marilyn A. Mendoza, "Terminal Lucidity Revisited: The Mystery Continues," *Psychology Today*, September 30, 2019, psychologytoday.com.

p. 194, **"Do what you want. I no feel good."** Oral Deposition of John Gnora Sr., Taken March 6, 1951, in the Case of John Gnora, Individually and as

Administrator of the Estate of Suzanne Gnora vs. American Steel & Wire Company, Civil Action No. 8077, Pennsylvania Western District Court.

p. 194, "**[She] just couldn't breath . . .** " Oral Deposition of John Gnora.

p. 194, "**He gave her an injection . . .**" Schrenk et al., *Air Pollution in Donora*, 50.

p. 194, "**A doctor arrived at some point . . .**" Schrenk et al., *Air Pollution in Donora*, 51.

p. 194, "**Born in 1882 in Poland . . .**" Schrenk et al., *Air Pollution in Donora*, 53.

p. 195, "**He died in his son's arms.**" "Father Collapses, Dies in Son's Arms: Five Orphaned by Donora Smog," *Pittsburgh Press*, November 1, 1948, 2.

26: Donora Goes to Press

p. 196, "**Then another man came in . . .**" Edwin Kiester Jr., "A Darkness in Donora," *Smithsonian Magazine*, November 1999, smithsonianmag.com.

p. 196, "**Neale at that point . . .**" "Donora Will Honor Neale, Industrialist," newspaper article provided by Lester Kilduff, Neale's grandson, February 2019.

p. 197, "**Chambon was on the phone . . .**" Roueché, "The Fog," 308.

p. 197, "**He referred to weddings as . . .**" Bernard Weinraub, "He Turned Gossip into Tawdry Power; Walter Winchell, Who Climbed High and Fell Far, Still Scintillates," *New York Times*, November 18, 1998, E-1; "Walter Winchell: The Power of Gossip," *American Masters*, September 15, 2020, www.pbs.org.

p. 197, "**Winchell's radio shows famously started . . .**" "Walter Winchell: The Power of Gossip."

p. 198, "**Stanley Tucci's line in the 1998 made-for-television movie . . .**" Movie Quotes, moviequotes.com.

p. 199, "**People dropped dead . . .**" Davis, "Heavy Air of Donora, Pa.".

p. 199, "**Soon as we got them above the smog . . .**" Karen Ivory, *Pennsylvania Disasters: True Stories of Tragedy and Survival* (Guilford, CT: Insiders' Guide, 2007), 98.

p. 200, "**Facial nerve injury was a common complication . . .** " Emails from Stephen Mass, MD, November 2020.

p. 200, "**Walter was put on a pedestal . . .** " Email from Nancy Rongaus Cherney, October 19, 2020.

p. 200, "**Walter served as a frontline surgeon . . .**" Interview with David Sommerville Sr., October 25, 2020.

p. 201, "**I picked up my grip . . .**" Roueché, "The Fog," 309.

p. 201, **"I knew then that we'd seen the worst of it . . . "** Roueché, "The Fog," 309.

p. 202, **"Blough called Neale . . ."** Snyder, "Death-Dealing Smog," 29.

p. 202, **"A Monongahela physician . . ."** "Donora: The 1948 Poisonous Smog That Killed 20 Led to Creation of the Clean Air Act," Fluoride Action Network, http://fluoridealert.org/.

p. 204, **"The greater part of the fog . . . "** Snyder, "Death-Dealing Smog," 29–30.

p. 204, **"Chambon told the assembled . . ."** Snyder, "Death-Dealing Smog," 30.

p. 205, **"He said that US Steel had banked . . ."** Snyder, "Death-Dealing Smog," 31.

27: Bless the Rains

p. 208, **"He didn't respond to any treatment . . ."** Schrenk et al., *Air Pollution in Donora*, 53.

p. 209, **"They are silent killers."** Scott Beveridge, "Panic, Condemnation and Redemption Follow the Deadly 1948 Donora Smog," *Travel with a Beveridge* (blog), http://scottbeveridge.blogspot.com/.

p. 210, **"You could breathe."** Roueché, "The Fog," 309–10.

p. 210, **"The smoke from the Zinc Works . . . "** Snyder, "Death-Dealing Smog," 31.

p. 211, **"When the rain stopped . . ."** Davidson, "Donora," 60.

p. 211, **"Hospitals in Monongahela and Pittsburgh . . ."** "Smog Patients Leaving Local Hospital; 10 Patients Treated," *Daily Republican*, November 1, 1948, 1.

p. 211, **"Even the trees in the cemetery . . ."** Roueché, "The Fog," 312.

p. 212, **"One such expert, Duncan A. Holaday . . ."** Edwin F. Brennan, "Zinc Plant Absolved in Deaths," *Pittsburgh Post-Gazette*, November 5, 1948, 1.

p. 213, **"He was, in fact, a leader in that field."** Victor E. Archer and Bobby F. Craft, "A Tribute to Duncan A. Holaday," *Applied Occupational and Environmental Hygiene* 5, no. 6 (2011): 390, https://doi.org/10.1080/1047322X.1990.10389660/.

p. 213, **"In his role as director . . ."** "Industrial Health," *Indiana Evening Gazette*, April 6, 1943, 4.

p. 213, **" . . . issued many warnings . . ."** "Parathion Danger," *Indiana Evening Gazette*, August 25, 1949, 13.

p. 214, **"Purdue wrote a letter of support . . ."** "Letter from Austin Purdue,

Bishop of Pittsburgh, to Mr. Neale," November 2, 1968, obtained from Les Kilduff, February 22, 2019, 1.

p. 214, **"Angelo F. Natali, a resident of Webster . . ."** Postcard from Angelo F. Natali to William Rongaus, November 1, 1948, courtesy of Nancy Rongaus Cherney.

p. 214, **"They were afraid of losing their jobs."** David Warner, "Donora Recalls Smog That Turned Killer 30 Years Ago," *Pittsburgh Post-Gazette*, October 23, 1978, 2.

p. 215, **"They were realistic . . ."** Email from David Lonich, November 2, 2020.

28: The Blaming Game

p. 218, **"Fluorine was considered, for a time . . ."** Bryson, *Fluoride Deception*, 129.

p. 218, **"Sadtler had graduated . . ."** Linda Wang, "Preserving a Legacy," *Chemical and Engineering News*, August 4, 2008, https://pubsapp.acs.org/.

p. 218, **"His father and, especially, his grandfather . . ."** "Portrait of Samuel Phillip Sadtler (1847–1923)," Science History Institute, https://digital.sciencehistory.org/.

p. 218, **"After those paragraphs, the newly degreed Sadtler . . ."** Letter from Philip Sadtler to Walter Rongaus, November 4, 1948, courtesy of Nancy Rongaus Cherney.

p. 219, **"Coal stoves being used to heat . . ."** Schrenk et al., *Air Pollution in Donora*, 164.

p. 219, **"The recommendation read, 'Establish a program . . .'"** Schrenk et al., *Air Pollution in Donora*, 165.

p. 219, **"Hood, who would be named president . . ."** "Walter F. Munford Named U.S. Steel President, Succeeds Clifford Hood," *Daily Courier*, May 8, 1959, 12.

p. 219, **"It is our desire . . ."** Maureen Gothlin, "PHS Report on Donora Smog Warns Other Industrial Towns," *Daily Republican*, October 14, 1949, 8.

p. 219, **"He claimed in the report's conclusion . . ."** Schrenk et al., *Air Pollution in Donora*, 162.

p. 220, **"Townsend was much more direct . . ."** Schrenk et al., *Air Pollution in Donora*, 162.

p. 221, **"Our stacks emit the same gases . . ."** Lynne Page Snyder, "The Death-Dealing Smog over Donora, Pennsylvania: Industrial Air Pollution, Public Health Policy, and the Politics of Expertise, 1948–1949," *Environmental History Review*, 18, no. 1 (Spring 1994): 137.

p. 221, **"Mills, a pioneering environmentalist . . ."** Clarence A. Mills, "The Donora Episode," *Science*, January 20, 1950, 67.

p. 221, **"Drinker also claimed, without evidence . . ."** Leslie Silverman and Philip Drinker, "The Donora Episode—A Reply to Clarence A. Mills," *Science*, July 21, 1950, 93.

p. 221, **"In December 1953 she presented a paper . . ."** Daniel Costa and Terry Gordon, "Mary O. Amdur," *Toxicological Sciences* 56, no. 1 (July 2000): 6, https://doi.org/10.1093/toxsci/56.1.5.

p. 222, **"Stepping close to her . . ."** Bob Musil, "New Alzheimer's Clues: Thank Mary Amdur?" Rachel Carson Council, January 1, 2017, https://rachelcarsoncouncil.org/new-alzheimers-clues-thank-mary-amdur.

p. 222, **"She refused, and Drinker fired her."** Costa and Gordon, "Mary O. Amdur," 6.

p. 222, **"He soon authorized a first-of-its-kind meeting . . ."** Harry S. Truman, "Message to the United States Technical Conference on Air Pollution," remarks delivered May 3, 1950, to the United States Technical Conference on Air Pollution, Washington, DC, www.trumanlibrary.gov.

p. 222, **"Truman headlined the conference . . ."** Truman, "Message to the United States Technical Conference."

p. 223, **"Ruckelshaus was just thirty-eight years old . . ."** "William Ruckelshaus, Who Quit in 'Saturday Night Massacre,' Dies at 87," *New York Times*, November 27, 2019, A29.

29: Fighting the Good Fight

p. 225, **"An auto-freight clerk named Leroy C. Le Gwin . . ."** Stephen Fletcher, "Donorians and the Good WILLmington Mission," *A View to Hugh* (blog), April 5, 2011, https://blogs.lib.unc.edu/.

p. 225, **"Broadfoot had been a heroic World War II fighter pilot . . ."** "Obituaries," *News and Record*, June 22, 2000, https://greensboro.com/.

p. 226, **"In the process he had become ill . . ."** "Forty Residents of Donora Begin Free Week-Long Vacation in N.C. to Rest Lungs in Beach Salt Air," *Daily Republican*, November 119, 1948, 6.

p. 226, **"Twenty-three men and seventeen women . . ."** "Donora Victims on Trip Listed: With Few Exceptions All Are Elderly," *Pittsburgh Press*, November 18, 1948, 2.

p. 226, " . . . under the auspices of Regina Dougert . . ." "Donora Smog Victims," photo, November 19, 1948, New Hanover County Public Library.

p. 226, "Ward, an immigrant from Wales . . ." James W. Ross, "Donora Smog
Victims Get Royal Welcome in South," *Pittsburgh Post-Gazette*, November
9, 1948, 14.

p. 227, " . . . even though US Steel knew about those chemicals . . ." *Gnora v.
American Steel & Wire*, Plaintiff's Amendment to Complaint, US District
Court for the Western District of Pennsylvania, Civil Action 8077, May 9,
1951.

p. 228, " . . . Kenworthey continued in his other defense. . ." *Gnora v. American
Steel & Wire*, Answer to Complaint, US District Court for the Western
District of Pennsylvania, Civil Action 8077, May 9, 1951.

p. 229, "For himself John Gnora was left . . ." *Gnora v. American Steel & Wire*,
Plaintiff's Amendment to Complaint, 6.

p. 229, "Like Gliwa and Pinkasiewicz . . ." "1948 Smog," Donora Historical
Society and Smog Museum, www.sites.google.com/site/donorahistorical
society/.

p. 229, "All were settled out of court . . ." Bill Kovarik, "Air Pollution," *Environmental History*, https://environmentalhistory.org/.

p. 230, "Although the Donora data do point . . . " Antonio Ciocco and Donovan J. Thompson, "A Follow-Up of Donora Ten Years After: Methodology
and Findings," *Journal of Public Health*, February 1961, 155.

p. 230, "In 2016 a ten-year-long study . . ." Joel D. Kaufman, Sara D. Adar, R.
Graham Barr, Matthew Budoff, Gregory L. Burke, Cynthia L. Curl, Mar-
tha L. Daviglus, Ana V. Diez Roux, Amanda J. Gassett, David R. Jacobs Jr.,
Richard Kronmal, Timothy V. Larson, Ana Navas-Acien, Casey Olives,
Paul D. Sampson, Lianne Sheppard, David S. Siscovick, James H. Stein,
Adam A. Szpiro, and Karol E. Watson, "Association between Air Pollution
and Coronary Artery Calcification within Six Metropolitan Areas in the
USA (The Multi-ethnic Study of Atherosclerosis and Air Pollution): A
Longitudinal Cohort Study," *Lancet*, August 13–19, 2016, 696.

p. 230, "Together those two conditions. . ." Hannah Ritchie and Max Roser,
"Causes of Death," Our World in Data, December 2019, https://
ourworldindata.org/causes-of-death.

p. 230, " . . . including low birth weight," Reihaneh Sarizadeh, Maryam Das-
toorpoor, Gholamreza Goudarzi, and Masoumeh Simbar, "The Associa-

tion Between Air Pollution and Low Birth Weight and Preterm Labor in Ahvaz, Iran," *International Journal of Women's Health* 12 (2020): 313.

p. 230, "… delays in the development …," Sarah Morton, Trenton Honda, Emily Zimmerman, Kipruto Kirwa, Gredia Huerta-Montanez, Alaina Martens, Morgan Hines, Martha Ondras, Ki-Do Eum, Jose F. Cordero, Akram Alshawabekeh, and Helen H. Suh, "Non-Nutritive Suck And Airborne Metal Exposures Among Puerto Rican Infants," *Science of the Total Environment*, May 26, 2021, https://doi.org/10.1016/j.scitotenv.2021.148008.

p. 230, "a worsening of diabetes," Chris C. Lim, Richard B. Hayes, Jiyoung Ahn, Yongzhao Shao, Debra T. Silverman, Rena R. Jones, Cynthia Garcia, and George D. Thurston, "Association between Long-Term Exposure to Ambient Air Pollution and Diabetes Mortality in the US," *Environmental Research*, August 2018, 330.

p. 230, "numerous chronic respiratory diseases," GBD Chronic Respiratory Disease Collaborators, "Prevalence and Attributable Health Burden of Chronic Respiratory Diseases, 1990–2017: A Systematic Analysis for the Global Burden of Disease Study 2017," *Lancet*, June 2020, 585.

p. 231, "a variety of cancers," "Air Pollution May Be Associated with Many Kinds of Cancer," American Association for Cancer Research, www.aacr.org/patients-caregivers/progress-against-cancer/air-pollution-associated-cancer.

p. 231, "A team of researchers at . . ." Mahdieh Danesh Yazdi, Yan Wang, Qian Di, Yaguang Wei , Weeberb J. Requia, Liuhua Shi, Matthew Benjamin Sabath, Francesca Dominici, Brent A. Coull, John S. Evans , Petros Koutrakis, and Joel D. Schwartz, "Long-Term Association of Air Pollution and Hospital Admissions Among Medicare Participants Using a Doubly Robust Additive Model," *Circulation*, February 22, 2021, 1584.

p. 231, "The study's lead author … " Jane E. Brody, "Air Pollution's Invisible Toll on Your Health," *New York Times*, June 28, 2021, nytimes.com.

p. 231, "Charlton remembered former Donora mayor John Lignelli . . ." Email from Brian Charlton, November 5, 2020.

p. 231, "The amendment expanded the act's focus . . ." "1990 Clean Air Act Amendment Summary," US Environmental Protection Agency, www.epa.gov.

Epilogue

p. 233, **"The steel mills made it another decade..."** "Steel Mill," *Donora Historical Society*, www.sites.google.com/site/donorahistoricalsociety/.

p. 233, **"More than 4,800 Pennsylvanians died..."** Logan Hullinger, "Pa. Leads Nation in Per Capita Premature Deaths Due to Air Pollution, Study Finds," *York Dispatch*, February 13, 2020, yorkdispatch.com.

p. 234, **"Pennsylvania's air quality isn't the nation's worst..."** Jordan Rosenfeld, "The Top Seven U.S. States with the Worst Air Quality," *Molekule*, October 16, 2019, https://molekule.science/.

p. 234, **"People living in Rome, Sarajevo..."** "Worldwide Air Quality: Air Quality Rankings," World Air Quality Index Project, https://aqicn.org/.

p. 234, **"I make a mental note..."** Furkan Latif Khan, "What It's Like to Breathe Some of the Most Polluted Air in the World," *NPR News*, November 25, 2018, www.npr.org.

p. 235, **"A city hardened by war..."** Kate Winkler Dawson, *Death in the Air: The True Story of a Serial Killer, the Great London Smog, and the Strangling of a City* (New York: Hachette, 2017), 158.

p. 235, **"Dawson explained that in London..."** Alessandra Potenza, "In 1952 London, 12,000 People Died from Smog—Here's Why That Matters Now," *The Verge*, December 16, 2017, theverge.com.

p. 235, **"Even without a museum..."** Christopher Klein, "The Great Smog of 1952," History.com, December 2, 2012.

p. 235, **"Since 1935 the average concentration..."** Hannah Ritchie, "What the History of London's Air Pollution Can Tell Us about the Future of Today's Growing Megacities," Our World in Data, June 20, 2017, https://ourworldindata.org/.

p. 236, **"After bulldozing an entire town..."** Morgan, *Western New York Steel*, 9.

p. 236, **"Wright made some pretext to take a biopsy..."** Donner, *Autobiography*, 126. The primary tumor was not located until an autopsy was performed.

p. 236, **"Joseph succumbed to his disease..."** "Joseph Donner, Son-In-Law of Eltings, Is Dead," *Chicago Tribune*, November 10, 1929, 16.

p. 237, **"An autopsy showed..."** Donner, *Autobiography*, 126.

p. 237, **"Donner wrote only..."** Donner, *Autobiography*, 126.

p. 237, **"Elizabeth charged Elliott with 'extreme cruelty'..."** "Divorced, Son of F. D. Flies to 'New Deal,'" *Daily News*, July 18, 1933, 3.

p. 237, **"Five days later Elliott married ..."** "Milestones, March 27, 1944," *Time*, March 27, 1944.

p. 237, **"He sold Donner Steel Company ..."** Donner, *Autobiography*, 128.

p. 237, **". . . established the International Cancer Research Foundation . . ."** "Cancer Research," *Nature*, June 16, 1934, https://doi.org/10.1038/133905a0.

p. 238, **"... and headquartered in Tarrytown, New York."** Email from Deirdre Feeney, August 30, 2021.

p. 238, **"He flew to Montreal ..."** "Philanthropist W. H. Donner Dies Aged 89," *Gazette*, November 5, 1953, 31.

p. 238, **"Zinc mill superintendent ..."** "Home of Champions: Honoring Milton Mercer Neale," Distinguished Citizens Club of Donora, April 24, 1963, luncheon program printed by Monongahela Publishing.

p. 239, **"He would've died ... "** Email from Lynn Harris, September 7, 2021.

p. 239, **"Feeling somewhat stronger . . ."** Sally E. Roth, "The Happy Time," unpublished essay provided by Jerry Harris, October 3, 2020, 2.

p. 239, **"In 1983 he received . . ."** "Rongaus Draws Top MVH Honor," *Valley Independent*, May 10, 1983, 1.

p. 240, **" ... a 1947 American LaFrance ... "** David Krumboltz, "Me & My Car: 1947 American LaFrance Fire Engine Found Home in Danville," *Mercury-News*, February 2, 2021, www.mercurynews.com.

Appendix: Complete Victim Data

p. 248, **"Fourth, a victim must have died ..."** Emails from George Leikauf, January 29, 2020.

SELECTED REFERENCES

Bleiwas, Donald I., and Carl DiFrancesco. "Historical Zinc Smelting in New Jersey, Pennsylvania, Virginia, West Virginia, and Washington, D.C., with Estimates of Atmospheric Zinc Emissions and Other Materials." Open-File Report 2010-1131, US Department of the Interior and US Geological Survey.

Brimblecombe, Peter, ed. *Air Pollution Episodes*. Singapore: World Scientific, 2017.

Cannadine, David. *Mellon: An American Life*. New York: Vintage, 2006.

Charlton, Brian. "Cement City: Thomas Edison's Experiment with Worker's Housing in Donora." *Western Pennsylvania History* 96, no. 3 (Fall 2013): 34–45.

Courland, Robert, and Dennis Smith. *Concrete Planet: The Strange and Fascinating Story of the World's Most Common Man-Made Material*. Buffalo: Prometheus, 2011.

Davidson, Bill. "Our Poisoned Air." *Collier's*, October 23, 1948.

Davidson, Bill. "Donora: The Case of the Poisoned Air." *Collier's*, October 22, 1949.

Eastman, Crystal. *Work-Accidents and the Law*. Chicago: Russell Sage Foundation, 1910.

Ivory, Karen. *Pennsylvania Disasters: True Stories of Tragedy and Survival*. Guilford, CT: Insiders' Guide, 2007.

Kiester, Edwin, Jr. "A Darkness in Donora." *Smithsonian*, November 1999. www.smithsonianmag.com/history/a-darkness-in-donora-174128118.

Parker, Arthur. *The Monongahela: River of Dreams, River of Sweat*. University Park: Pennsylvania State University Press, 1999.

Peterson, Michael. "Thomas Edison's Concrete Houses." *American Heritage Invention and Technology* 11, no. 3 (1996).

Ross, Benjamin, and Steven Amter. "Donora's Strangler Smog." In *The Polluters: The Making of Our Chemically Altered Environment,* by Benjamin Ross and Steven Amter, 86–97. New York: Oxford University Press, 2010.

Roueché, Berton. "The Fog." *New Yorker,* September 30, 1950.

Schrenk, H. H., Harry Heimann, George D. Clayton, and W. M. Gafafer. *Air Pollution in Donora, Pa: Epidemiology of the Unusual Smog Episode of October 1948: Preliminary Report.* Washington, DC: US Public Health Service, 1949.

Snyder, Lynne Page. "The Death-Dealing Smog over Donora, Pennsylvania: Industrial Air Pollution, Public Health, and Federal Policy, 1915–1963." PhD dissertation, University of Pennsylvania, 1994.

Stacey, Charles E., Brian Charlton, and David Lonich. *Images of America: Donora.* Charleston: Arcadia, 2010.

Vecsey, George. *Stan Musial: An American Life.* New York: Random House, 2011.

Walker, Charles Rumford. *Steel: The Diary of a Furnace Worker.* Boston: Atlantic Monthly, 1922.

Websites of Interest

Beveridge, Scott. *Travel with a Beveridge* (blog). http://scottbeveridge.blogspot.com.

Donora Historical Society. www.sites.google.com/site/donorahistoricalsociety.

"Dumping Slag at Bethlehem Steel in 1994." YouTube. www.youtube.com/watch?v=zhJF_hTJ2Rw.

"Zinc Works at Swansea." YouTube. www.youtube.com/watch?v=IdvTCz65W5U.

INDEX

Note: Page references in *italics* refer to figures and tables.

INDEX

INDEX